Mitchell Masterpieces

AN ILLUSTRATED HISTORY OF B-25 WARBIRDS IN BUSINESS

VOLUME 3

Mitchell Masterpieces

AN ILLUSTRATED HISTORY OF B-25 WARBIRDS IN BUSINESS

VOLUME 3

WIM NIJENHUIS

Lanasta

Eleven B-25s at a reunion on 17-18 April, 2017 at Wright-Patterson AFB, Dayton, Ohio. It was the 75th Anniversary reunion of the Doolittle Tokyo Raiders. The B-25s from across the United States were in attendance to honour the raiders. The anniversary was attended by the sole surviving member of the Doolittle Raiders Lt. Richard E. Cole. The airplanes are all in airworthy condition with beautiful nose art, mostly derived from the original art displayed by their counterparts in World War II. (Dennis Nijenhuis)

ISBN: 978-94-6456-066-4
E-ISBN: 978-94-6456-067-1

NUGI: 465
1st print, January 2024

© Copyright 2024
Walburg Pers/Lanasta

Print: PrintSupport4U
Design: Jantinus Mulder

www.walburgpers.nl/lanasta

All correspondence regarding copyrights, translation or any other matter can be directed to: Lanasta, Nieuwe Prinsengracht 89, 1018 VR Amsterdam, the Netherlands.

Contents

FOREWORD

By Lt-Gen (retired) Bart Hoitink,
Former chairman of the Royal Netherlands Air Force Historical Flight

THE SOUND OF FREEDOM!

Sometimes you meet special people. Ordinary people who make a choice under difficult circumstances. A choice that will determine the rest of their lives. A choice that has contributed to our freedom today. Some people resisted the oppression during World War Two. Others decided to take to the skies to work for our freedom. They chose a dangerous mission. Many airmen did not return from their missions over occupied territory.

Among the many aircraft deployed during the war and the liberation were B-25 Mitchell bombers and Spitfires fighters. These aircraft were part of Dutch units of the RAF No. 320 and 322 Squadron respectively, and two of them now of the Royal Netherlands Air Force Historical Flight.

One special aircraft catches the eye. Our B-25N Mitchell named "Sarinah". The Mitchell was employed by the ML KNIL, the Military Aviation Department of the Royal Netherlands East Indies Army. In addition, the Mitchell was used in No. 320 (Dutch) Squadron in the Royal Air Force. After the war, the Mitchell served in the Netherlands East Indies during the police actions. After the hostilities, most of the aircraft were handed over to the Indonesian Air Force. In our country, the Mitchell was used by the Naval Aviation Service until 1954 with No. 320 Squadron that was discontinued in 2005. The aircraft mainly served as a Search-And-Rescue aircraft, including during the flood disaster of February 1953.

The B-25 in service with the Royal Netherlands Air Force Historical Flight has an American history. The aircraft was in service with the USAF from 1944-1959 and then, through various owners in America, finally sold in 1990 to the Duke of Brabant Air Force (DBAF) in the Netherlands. Ultimately, the aircraft was transferred from the DBAF to the Royal Netherlands Air Force Historical Flight in 2010. Meanwhile, in 1999 the colour scheme was adapted to the ML-KNIL period and the aircraft was named "Sarinah".

The sound of the engines gave hope to the people on the ground, soldiers, and civilians. Hope for a speedy release. Hope for a better life. The sound of the aircraft engines was music to the ears. And at the Royal Netherlands Air Force Historical Flight at Gilze-Rijen Air Base, we go back in time and regularly listen to this "music".

A Mitchell bomber, Spitfire, Harvards, Ryan, Piper Cubs, Tiger Moth, and many other aircraft, which have played an important role in Dutch aviation history, are kept in perfect condition by many volunteers.

Special people, but also special aircraft, have fought our freedom. The stories must not be lost, the music must not stop!

Sound of Freedom.

ACKNOWLEDGEMENTS

This book would not have been possible without the generous help of other people. Many sources were consulted, and individuals or organisations allowed me the use of their photographs or other material.

I am happy that Bart Hoitink wanted to write the foreword in this book. Bart Hoitink, Lieutenant General (retired) at the Royal Netherlands Air Force, was chairman of the Royal Netherlands Air Force Historical Flight. This organisation operates an impressive fleet of historical aircraft including her flagship "Sarinah", a B-25 which previously flew in in the U.S.A. under the name "Cochise". Bart Hoitink was, among other things, helicopter pilot, squadron commander and Inspector General of the Netherlands Armed Forces. He is a great advocate of preserving historic aircraft and keeping this heritage flying. This, however, always under strict safety requirements. Thank you, Bart, for writing the foreword.

A word of thanks goes to Scott A. Thompson from the Sacramento Area, California, and his particularly useful B-25 information and pictures of Aero Vintage Books on Aerovintage.com. Scott Thompson has long held an interest in the B-25 and other warbirds. He started collecting information and photos on these warbirds when he was a teenager and began writing about them in 1980. He has written numerous articles and books, including the beautiful and informative book "B-25 Mitchell in Civil Service".

I want to mention Hélio Higuchi from Brazil. He is researching Latin American military history facts and equipment. He put me in touch with aircraft historians Tony Sapienza in Paraguay, Gary G. Khun and John Davis from the U.S.A., and Claudio Cáceres Godoy, one of the greatest enthusiasts of Chilean aviation history. All five historians provided me with remarkably interesting material and/or pictures. Therefore, many thanks.

I would also like to thank the following individuals or organisations (in alphabetical order) who have helped me, in the past or recently, for their time and effort for providing information or pictures that could be used in this book.

★ Bob Hobbs	United Kingdom
★ Bob Woolnough	United Kingdom
★ Canadian Warplane Heritage Museum	Canada
★ Carl Geust	Finland
★ Cor van Gent	Netherlands
★ Dan Hagedorn	U.S.A.
★ Dennis Nijenhuis	Netherlands
★ Fred Pelder	Netherlands
★ Gary "Olde Carl" Schenauer	U.S.A.
★ Gary Vincent	Canada
★ Ingo Warnecke	Germany
★ Malcolm Taylor	U.S.A.
★ Michael Baldock	United Kingdom
★ Neil Aird	Canada
★ Norman L. Avery	U.S.A.
★ (former) North American Rockwell	U.S.A.
★ Proyecto B-25 Mitchell (Huaira Bajo)	Argentina
★ Richard Vandervord	United Kingdom
★ Rodney Small	United Kingdom
★ Steve Fitzgerald	United Kingdom
★ San Diego Air & Space Museum	U.S.A.

A word of appreciation and thanks is in place to the people behind reliable and informative public sources/websites from which useful information could be obtained. Particularly useful sources were:
★ Aerial Visuals of Mike Henniger, Canada
★ B-25history.org of Dan Desko, U.S.A.
★ Federal Aviation Administration of the United States Department of Transportation. U.S.A.
★ The Internet Movie Plane Database
★ Warbirds Directory of Geoff Goodall, Australia
★ Warbirdregistry.org of Warbirds Resource Group, U.S.A.
★ www.napoleon130.tripod.com of Martin J. Simpson, U.K.

Finally, I want to thank Rob van Oosterzee from the Netherlands, for his great and conscientious help with the linguistic issues of the manuscript, and for providing some useful suggestions.

Photo courtesy: Where known, the photo sources are mentioned.

INTRODUCTION

In June 1978, for the first time in my life I saw a flying B-25 at Deelen Air Force Base during the 65th Anniversary of the Royal Netherlands Air Force. This airplane was the civil registered "Gorgeous George-Ann" and was provided with a very nice nose art. Three years later in September 1981, a second B-25 made an air show at Soesterberg Air Force Base. This plane was named "Big Bad Bonnie". Both airplanes were owned by John Hawke and were used in the movie "Hanover Street". At that time, I could not suspect that about 40 years later these same B-25s would be a subject of my third part of "Mitchell Masterpieces". In the meantime, I have been a few times in the U.S. and have seen and photographed several B-25s. A highlight was my last trip in April 2017, with my son Dennis to the B-25 gathering at Grimes Field and Dayton, Ohio, on the occasion of the 75th Anniversary of the Doolittle Raiders. The photos taken during those trips, could be used well for this Volume 3. In this third and final book of "Mitchell Masterpieces" about the North American B-25 Mitchell in service, the B-25s will be described which were operated in civil service in all different countries. The other Volume 1 dealt with the B-25s in service with the U.S. Armed Forces and Volume 2 covered the B-25s in service with all the foreign military forces.

In 1939, the B-25 was developed as a medium bomber. Therefore, it had a primarily military role during the war. The B-25 bomber served in the armed forces of the United States as well as other allied countries. The major other allied countries were the United Kingdom and the Soviet Union and relatively large numbers of B-25s were also used by the Netherlands, Canada, China, Brazil and Venezuela. But also, some other countries used the B-25 after the war. Only after the war, the B-25 played a role in civil aviation. In the post-war period, various remaining aircraft were sold by the defence ministries. The major supplier was of course the U.S.A. Many aircraft were stripped of their armament and other military equipment and often reached the civilian market through auctions. They passed into private hands and were deployed for transportation of personnel or cargo, for training, firefighting, agricultural spraying, or other purposes. Even sometimes for smuggling. Some were converted into a luxury business plane.

Many B-25s that served after the war were overall natural aluminium finished. However, there were some colourful examples with or without company logos and other features. Far after the war, the so-called warbird circuit emerged. Old aircraft were refurbished and even made airworthy by historians and aviation enthusiasts. And nowadays, there is an increasing number of warbirds, including the B-25. They are often painted in fantasy colours, but sometimes as much as possible in the colours of their original predecessors at the time of the war. Like the two previous volumes of "Mitchell Masterpieces", this book by no means pretends to be complete. The subject is by far too complex and detailed to be described in full here. It is impossible to describe and illustrate all the companies and individuals who used the B-25 and the history, colours, and markings of all the civilian B-25s in one book. Moreover, the information on many airplanes is no longer available and often remains a mystery. And sometimes, photographs are scarce or unfortunately of poor quality. Therefore, choices had to be made. There is no plane-by-plane chronological overview. That is why in the different chapters several times the same airplanes are described, but always tailored to the period in which they were employed by the various companies or organisations. Where necessary, the companies as well as museums have been put in alphabetical order as much as possible. This book should serve as a general view of the companies, organisations, and owners and the B-25s they flew. Moreover, the book would not be complete if no attention was paid to the civilian B-25s in the various museums, film industry and the warbird circuit. That is why this has been also described in detail.

Together with Volume 1 and Volume 2 of "Mitchell Masterpieces", the three books give an excellent overall view of the use and service of the B-25 in the various countries and the colours and markings in which they were operated.

Wim Nijenhuis

SCRAPPING AND STORAGE

The United States produced approximately 300,000 aircraft for World War Two. By 1944 the U.S. Foreign Economic Administration began a program to scrap certain obsolete, damaged, and surplus military aircraft overseas. After the war, many of the U.S. military aircraft overseas were not worth the time or money to bring back the airplanes to the United States and so they ended on scrap yards on the battlefield. They were consequently buried, bulldozed, or even sunk at sea.

Most, however, were returned home for storage, sale, or scrapping. From 1945, thousands of American aircraft began arriving back in the United States. The country had a huge surplus of aircraft and they were turned over to the Reconstruction Finance Corporation (RFC). The RFC and the War Assets Administration (WAA) handled the disposal of these aircraft. The RFC was a government corporation in the United States between 1932 and 1957 that provided financial support to state and local governments and made loans to banks, railroads, mortgage associations and other businesses. Its purpose was to boost the country's confidence and help banks resume daily business activities after the start of the Great Depression. The RFC established depots around the country to store and sell surplus aircraft. By the summer of 1945, at least 30 sales-storage depots and 23 sales

A late B–25D model of the 42nd Bombardment Group "The Crusaders" upside down on a scrap yard somewhere in the Pacific area. (USAF)

centres were in operation. One of the largest disposal facilities was at Kingman Army Airfield, Arizona. This location, also known as Storage Depot #41, was chosen because of its huge open spaces, good weather for aircraft storage, and three runways, one of which was 6,800 feet in length. Kingman was a field for short-term storage and subsequent disposal. At Kingman, many of the

old planes were sold to companies which cannibalized them for parts and turned the material to other commercial uses than aviation. Before the planes were offered for sale they were stripped of all confidential equipment, such as bomb sights, radar, and some radio installations. But most of the airplanes were scrapped. After a study, it was established that too many man-hours were re-

Kingman was not the only place where B-25s were stored. On this 1946 picture, at least twelve surplus B-25s are parked on the RFC storage depot at Ontario, California. There are also many B-26s and B-17s. (Bill Larkins)

quired to dismantle the airplanes for parts, and the cost of storage areas for the parts was too high. So, the method of salvage and melt was adopted. Main components such as engines, armament, instruments, and radios were removed from each airplane. The remainder of the aircraft was cut into pieces, and pushed into a large furnace, or smelter. Aluminium was the prime metal sought

after, melted, and poured into ingots for sale and shipping. The first aircraft arrived at Storage Depot #41 on 10 October, 1945. By December of 1945, every few minutes an aircraft arrived. Some were flown by their own combat crews directly from overseas. The veteran warbirds did arrive with bullet-ridden fuselages, splintered propellers, and coughing engines representing dozens of

missions. At the other extreme were aircraft right off the assembly line, their only flying time being their trip to Kingman. By April, 1946, the "Arizona Graveyard Air Force," as it was dubbed by the press, presented an amazing sight of more than 7,000 bombers, fighters, and other aircraft lined up row on row for six and a half miles along Route 66. Most of the aircraft, nearly 5,500, were sold to the Wunderlich Contracting Company of Jefferson City, Missouri, owned by Martin Wunderlich which paid $ 2,780,000 for the aircraft at Kingman. These included 140 B-25s. He built a smelter near the airfield and melted the aircraft mainly for their aluminium content. Before he did that, however, he siphoned the fuel they contained, and it was said that Wunderlich made more from the sale of the fuel than he paid for all the airplanes. By the end of 1948, all aircraft had been smelted into aluminium ingots, along with their nose art paintings.

So, many airplanes including B-25s were salvaged and melted. Other planes were transferred to civilian operators, or to the air forces of allied countries. Remaining airplanes were classified as obsolete or eligible for the strategic aircraft reserve. The airplanes were then assigned to an airfield for short-term storage and subsequent

B-25s at Kingman, Arizona in March 1946. After the war, thousands of warbirds which fought in the skies over the several war theatres, were parked row on row at the old Kingman Army Airfield. As reported in the Los Angeles Times of 1 April, 1946 among the types of ships stored at the depot were 141 B-25 Mitchells with the standard price being asked for them by the WAA of $8,250 each. (Phil Bath)

disposal, or for longer-term storage. Therefore, Kingman was not the only place were B-25s were stored. They were also present at Ontario Surplus Depot at Cal-Aero Field, near Chino, California, at Altus Army Airfield in Oklahoma, at Searcy Field in Stillwater, Oklahoma, at Pyote Army Airfield, Texas, at Davis-Monthan Air Force Base, Tucson, Arizona and at Walnut Ridge Army Airfield, located in Northeast Arkansas. The war-weary B-25s were not only stored at these depots. At the end of the war, the federal government's RFC set up a depot in the Fairfax district to sell machinery, tools, and similar equipment to peacetime industries and the public and thereby partially recoup the government's war cost. Some of the B-25 bomber plant items were sent to the depot, while others were transferred for government use elsewhere. Such materials as aluminium sheets and steel products went to reclamation centres, and spare airplane parts were dispatched to maintenance depots. Seventy-two incomplete, but flyable B-25s were not completed and accepted contractually by the Army Air Forces. They were part of the Terminal Flyable Inventory at the end of the production in Kansas City and were flown directly to the lots of the RFC for disposal.

On V-J day, the Army Air Forces possessed nearly 2,500 B-25s. Most of the older B-25C, D, and H bombers were immediately released to the RFC for disposal, but many hundreds of newer B-25Js were retained. About 1,800 B-25s remained in U.S. service. By 1950, the U.S. Air Force still had about 1,400 B-25Js in usable condition. Many of those B-25s had to be converted for their new operational role. That was also the case with thousands of other warplanes. Many of them, including B-25s, were updated and refurbished by the Grand Central Aircraft Company at Grand Central Airport, Glendale, California. These airplanes were then intended for the Nationalist Chinese Air Force, the Brazilian and Uruguayan Air Force, Standard Oil, Sinclair Oil and many airlines.

Other companies that modified many B-25s were Hayes Aircraft Company of Birmingham, Alabama and Hughes Tool Company of Culver City, California. Between 1951 and 1955, Hayes modified approximately 450 B-25Js for pilot training purposes and about 1,000 were overhauled by Hayes for IRAN (Inspect and Repair As Needed). Hughes of Culver City modified approximately 150 B-25Js for bombing and firefighting purposes. The major modifications varied through the years. They could include removal of armament and armour protection, modernising instrument panel and hydraulics, new windshields, modern heating systems, revising radio and interphone system, redesigned electrical systems, replacing the "S" stacks on the top seven cylinders of engines with an exhaust semi-collector ring, new carburettors with larger boxy air intake scoops, new oxygen systems, new heating, updating radar and fire control systems. On many airplanes, a specially designed black nose cap with a radome was mounted instead of the original transparent nose cap of the B-25J. Most of the modified B-25Js were redesignated as TB-25J. Later, several aircraft were designated as ETB-25J or JTB-25J (electronics development aircraft), VB-25J (staff transport) or CB-25J (utility aircraft) considering their operational role. For a short period of time, some B-25s were redesignated as AT-24 (advanced trainer). The modified TB-25 trainers were followed by a model letter. TB-25K was the designation given to trainers for the operators of the E-1 radar fire control system. Modified B-25Js for specialized advanced pilot training were designated TB-25L. The TB-25M was essentially the same as the K model except for the installation of the more advanced E-5 fire control system. The TB-25N was similar to the preceding TB-25L but was fitted with R-2600-29A engines. The USAF began to withdraw its last B-25s from service in 1957. Hundreds of airplanes were stored at Davis-Monthan and they became available for civil use.

The last 72 aircraft built were not completed and accepted contractually by the Army Air Forces. They were part of the Terminal Flyable Inventory at the end of the production in Kansas City and were flown directly to the lots of the RFC for disposal. On the photo a number of North American employees with the very last B-25 built, s/n 45–8899, prior to her departure from the Fairfax plant on 31 October, 1945.
(b–25history.org)

A 1960 announcement of the Redistribution & Marketing Division at Davis-Monthan of old WWII airplanes including 30 B-25s! At right, a 1977 advertisement of "Warbirds of the World" Inc. at San Marcos, Texas. N9494Z is the most expensive B-25 from the advertisement. In 1969, she flew in the movie "Catch-22" as "The Abominable Snow Man" and later as "Laden Maiden". In 1975, she was sold to John J. Stokes of "Warbirds of the World", based in San Marcos and flew in her Desert Sand colours also as "Laden Maiden". Currently, she is restored in Belgium. (Collection Wim Nijenhuis, Collection Doug Fisher)

THE CIVIL MARKET

The B-25 outnumbered all other medium bombers in USAF service and was soon converted into advanced pilot trainers after the war. A lot of B-25s were sent to the U.S. Air National Guard in support of F-89 Scorpion and F-94 Starfire fighter intercept squadrons. Others were utilised as trainers, weather reconnaissance and personnel transports. A few B-25s were even used in the Korean War as electronic warfare aircraft. But as said, after the war, also many B-25s were scrapped and melted. But fortunately, not all. Sale of military aircraft became possible through the Surplus Property Act of the Reconstruction Finance Corporation. The Surplus Property Act of 1944 was an act of the United States Congress that was enacted to provide for the disposal of surplus government property.

In the late 1950s, some 180 well-maintained ex-USAF B-25Js were stored at Davis Monthan, Arizona. Most of these were sold by auctions and passed into the hands of private U.S. owners or other countries. Sales in 1957 saw prices between $5,000 and $7,000 for a B-25. But later, prices dropped to between $1,500 and $2,000.
In the 1970s, when the airplanes began to be scarce, prices rose again. The civilian B-25s were mostly operated by companies all over the U.S.A. Civil B-25s were used for many purposes. After all military equipment had been removed, some were used

almost directly without modification. Others were completely rebuilt for special purposes. Sometimes the bomb bay was being faired over or plush interior accommodations were added. Some aircraft even had wingtip tanks.
They were used as personal transports, freight transports, agricultural sprayers and even as fire fighters. As aerial agricultural pesticide sprayers they were fitted with underwing or undertail booms for dispensing pesticides. The pesticide tanks were installed in the fuselage and wings as well as in the bomb bay. The U.S. Forest Service bought a substantial number of surplus B-25s, installed huge water tanks in the main fuselage and sent them out to waterbomb forest fires.

In the post-war years, about 70 companies spread across the U.S.A. flew the B-25 on air tanker and agricultural spraying duties. Several B-25s ended up being employed by the film industry, where in particular Tallmantz Aviation at Santa Ana, California, has played a large role. In the course of time, several of the civilian U.S. B-25s were sold to other countries. Especially in Latin American countries, various B-25s flew in civilian colours. However, there is less known and little information available about Latin American B-25s and their assignments and missions.

The civilian B-25s were mostly the later B-25J models and, therefore, in general overall natural aluminium finished. They were provided with civil registrations and sometimes the name of the company or organisation. Some, however, were completely repainted and we see examples with the most colourful designs. In civilian service, tail codes and tail art took on a whole new meaning with operators, be they air tanker or executive conversions, using the fin to display their own unique design that usually represented the company the aircraft flew for. Civil B-25s are still flying or have flown in nearly twenty different countries. In the next chapters the civil B-25s are further described for each country. But as mentioned before, this book by no means pretends to be complete. The subject is by far too complex and detailed to be described in full here. A general view of companies and services is given. Not every aircraft is described in detail and a choice has been made. In particular, the many U.S. airplanes and companies are not all described. Another problem is the airplanes of some other countries. Their civilian B-25s are not all known. Of many little is known, sometimes just rumours and sometimes nothing at all. So, these B-25s are very unclear and, unfortunately, there is no photographic material available. Despite these limitations, an attempt has been made to obtain a good overview of the civilian B-25s and, where necessary, the most colourful,

special, or remarkable airplanes are mentioned and illustrated. The volume also does not show the full history of the airplane, but it focuses mainly on her history and service in the particular country, organisation or company. Also, attention has been paid to the civilian B-25s in the various museums, film industry and the warbird circuit.

For reasons of readability of the book it was decided to mention as much as possible the civil aircraft numbers. For a comparison of the civil aircraft numbers with the U.S. Army Air Forces block and serial numbers mentioned in this book, see Appendix 1 at the back of the book.

U.S.A.

The U.S.A. is the country that used most of the B-25 airplanes produced. This applies to both military and civil aviation. As previous mentioned, after the war many military surplus B-25s were sold to private companies. They were often used as firefighting airplane or for agricultural spraying. Also, several B-25 airplanes were used in the American film industry. And finally, many ended up in a museum. Fortunately, there is still a reasonable number left that is airworthy, and today still fly around as highly respected warbirds. Many of these airplanes and companies are described and explained in the next chapters and as much as possible in alphabetical order.

Firefighting and Agricultural Duties

The United States Forest Service (USFS) is an agency within the U.S. Department of Agriculture. It has existed for more than one hundred years with the express purpose of managing public forests and grasslands. Starting in 1876, and undergoing a series of name changes, the USFS grew to protect and utilise millions of acres of forest on public land.

Federal forest management dates to 1876 when Congress created the office of Special Agent in the U.S. Department of Agriculture to assess the quality and conditions of forests in the United States.

In 1881, the Department expanded the office into the Division of Forestry. A decade later, Congress passed the Forest Reserve Act of 1891 authorising the President to designate public lands in the West into what were then called "forest reserves." Responsibility for these reserves fell under the Department of the Interior until 1905 when President Theodore Roosevelt transferred their care to the Department of Agriculture's new U.S. Forest Service.

Within the USFS, aviation is used for the rehabilitation of forests, search and rescue, law enforcement, personnel transport, aerial photography as well as fire prevention and suppression. The USFS, Fire and Aviation Management team lists this as their official stance on the use of aviation to combat fires. The primary mission of the Forest Service Aviation is to support the ground fire fighter through a variety of means including safe delivery of smokejumpers, propellers, air attack, fire fighter and cargo transport, surveillance, aerial reconnaissance and fire intelligence gathering, and aerial delivery of fire retardant and water.

Commercial aerial fire bombing began during the 1950s. The USFS had traditionally relied on contracting with private companies to provide large air tankers for fighting forest fires, the majority of which had been retired World War II and Korean War-era transports, bombers, and maritime patrol aircraft. By the end of the 1950s, various aircraft, mostly World War Two and Korean War surplus airframes, had found a new life as aerial fire fighter.

In the U.S.A., the airplanes and their methods used in aerial firefighting are generally called air tanker. In Canada, they also use the term water bomber. A considerable number of surplus B-25s found use with contractors who flew these aircraft for the USFS as fire-fighting aerial tankers. In the bomb bays huge water tanks or fire-retardant chemical tanks were installed. The B-25 Mitchells were used as fire tankers in the late 1950s and 1960s until banned by USFS, following a series of fatal crashes due to structural failure during the pitch-up after dropping their retardant load. The B-25 started being converted in 1959 and by 1960, for example there were sixteen B-25 tanker operators in the State of California. Most B-25s carried 1,000-gallon tanks either belly mounted or fitted internally in the bomb bay. Early success varied with each operator, however, in July of 1960 there were four fatal crashes within days of each other involving the B-25. Subsequently, the B-25 was banned from use in the State of California. Testing was conducted at Edwards Air Force Base in 1963 to determine the cause of certain flight control issues which resulted in the limited use of the B-25 elsewhere around the country and Canada.

The overall natural aluminium finished N6123C was used as an agricultural sprayer. Here she is seen around 1963, at the time she was owned by Abe's Aerial Service Inc. Note the large tank below the fuselage. (napoleon130.tripod.com)

ABE'S AERIAL SERVICE INC., SAFFORD, ARIZONA

Abraham (Abe) B. Sellards, Safford, Arizona operated an agricultural application business from Safford, Arizona from at least 1958. The airplanes of Abe's Aerial Service included two B-25s, N6123C and N9117Z. He has spent many months and thousands of dollars converting the surplus B-25s with tanks and fast-action belly doors. In early 1963, Abe Sellards merged with air tanker firm Aircraft Specialties Inc. at Mesa-Falcon Field Arizona, under a new parent company named Aviation Specialties Inc. Further description under Aviation Specialties Inc.

AERO FLIGHT INC., TROUTDALE, OREGON

Aero Flight Inc. was an air tanker business founded by Robert M. Sturges. He also was proprietor of Columbia Airmotive Inc. at Troutdale Airport. Sturges' passion was aviation and he did know a lot about the WWII B17 Flying Fortress bomber. He served as a technical representative for Boeing in England during World War Two. Sturges succeeded Glenn Otto as Mayor of Troutdale in 1972 and served until 1982. He designed liquid tanks for various ex-military aircraft, initially for agricultural spraying work. He also was involved in the very early fire-

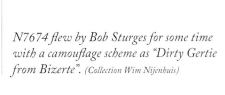

N7674 flew by Bob Sturges for some time with a camouflage scheme as "Dirty Gertie from Bizerte". (Collection Wim Nijenhuis)

bombing experimentation. Robert's son Ray Sturges was also involved with Aero Flight Inc. Through his business Columbia Airmotive, Bob Sturges acquired a variety of aircraft. The fleet of Aero Flight Inc. included among a few other airplanes, three B-17s, three P-40s, two A-26s and one B-25. This last mentioned was N7674, purchased in January 1969. This was a B-25J-25, s/n 44-30421, and was flown in camouflage colours and named "Dirty Gertie from Bizerte". In June 1978, the airplane was bought by Carl Scholl and Tony Ritzman of Historical Aircraft Preservation Group, Borrego Springs, California. The airplane was then parked at the Troutdale Airport. Scholl drove a replacement engine to Troutdale and got the bomber ferried back to Chino for restoration to airworthy condition. Around 1980, her life ended up in Colombia after a crash landing.

AERO INSECT CONTROL INC., RIO GRANDE, NEW JERSEY

Aero Insect Control Inc. was a pest control company spraying with over ten Boeing Stearmans and a Grumman TBM Avenger tanker test airplane. The company was located at Cape May County Airport, Rio Grande, New Jersey. By 1959, the principal was Maurice Curtis Young. In 1959, the company bought three B-25s.

N9076Z
One airplane was B-25J-25 with s/n 44-30772. She was stored at Davis Monthan in 1959 and bought by Maurice C. Young from Aero Insect Control Inc. She was registered as N9076Z. The company used her as tanker, and she flew as tanker #06C. In June 1964, she was sold to James L. Range, College Park, Georgia.

N9075Z
A second B-25 was 45-8896, a B-25J-35. After her service with the USAF, she was sold to Aero Insect Control Inc. in December 1960 and registered as N9075Z. In 1963, she was sold to Robert E. Messmer, Boca Raton, Florida.

N3514G
The third Mitchell from Aero Insect Control Inc. was a B-25J-30 with s/n 44-86786. Also stored at Davis Monthan and sold to Fogle Aircraft, Tucson, Arizona in July 1959. She was registered as N3514G. In November 1959, she was bought by Aero Insect Control. But already in June 1960, she was sold to Avionics Inc., Virginia Beach, Virginia. The airplane was out of service by 1970.

One of the three B-25s from Aero Insect Control Inc. in the early 1960s. This is N9076Z, the company used her as a tanker and flew her as #06C. The former USAF number BD-772 is still vaguely visible. (Collection Wim Nijenhuis)

AERO RETARDANT INC., FAIRBANKS, ALASKA

This tanker company was based at Fairbanks, Alaska in the early 1960s. By 1968, the company was owned by Don G. Gilbertson. He had been heading of this firefighting organisation since its beginning. He was raised in Fairbanks. He got his start in business as a truck driver for his dad and later he went to work for Alaska Freight Lines driving semi-trucks over the highway for four years while learning to fly. He then flew for Wien Airlines for 15 years and flew for Interior Airlines as chief pilot for a year and a half. By 1964, along with James Anderson, Gilbertson went into the firefighting business with a pair of World War II vintage airplanes,

N3514G somewhere in the 1960s, in the weeds at Friendship Airport, now Baltimore-Washington International Airport. (J.G. Handelman)

B-25s and a Cessna 180. Two years later Gilbertson bought out his partner and then incorporated in 1968 as Aero Retardant Inc. In 1972, the corporation's name was changed to Pacific Alaska Airlines. One of the most famous names in northern aviation, who was the forerunner of Pan American Airlines in Alaska. Aero Retardant Inc. had some DC-3s and DC-6s and two B-25s, s/n 44-86791 and s/n 44-86747.

N8196H

The B-25 with s/n 44-86791, registered as N8196H, was sold to Aero Retardant in February 1963. In June 1977, Pacific Alaska Airlines purchased her. In August 1979, she was sold to Donald Gilbertson. She was used in the television production of "Young Joe, The Forgotten Kennedy". In June 1982, he was sold to Aero Heritage Inc. of Melbourne, Australia.

N8163H

The other B-25 with s/n 44-86747 was registered as N8163H. Aero Retardant purchased her in April 1967. D. Gilbertson and James Anderson operated her as tanker #7. By February 1974, she was noted as derelict at the Fairbanks Airport in Alaska and in February 1977, she was sold to Noel Wien of Anchorage, Alaska.

N8163H probably in the late 1970s, prior to restoration. Her free stylized figure tanker #7 is still on her vertical tail. (b–25history.org)

AERO SPRAY INC., VANCOUVER, WASHINGTON

Aero Spray Inc. was an agricultural spraying operator for the fruit orchards of the Yakima Valley. The company bought two B-25s from the USAF disposals and fitted them with 1,000-gallon liquid tanks for fire bombing. The company's involvement in fire bombing seems to have ended in 1960 when both B-25 tankers were sold.

Ralph and Gini Richardson with their daughter in 1959 in front of N9866C.
(Fam Richardson)

N9865C

The other Mitchell was s/n 44-28834, also a B-25J-15. This airplane was also purchased from the USAF in October 1958 and registered as N9865C. In May 1960, she was converted into a fire tanker. But already in October 1960, she was sold to Western Air Industries, Redding, California.

N9865C stands on the field, apparently in reasonably new condition. The picture was probably taken in the late 1950s. (Collection Wim Nijenhuis)

AERO UNION CORP., CHICO, CALIFORNIA

N9866C

One Mitchell was s/n 44-28833. After storage at Davis Monthan, she was purchased by Aero Spray in October 1958. She was registered as N9866C. In May 1959, she was converted to a fire bomber, fitted with a 1,000-gallon tank. The airplane was operated for the 1959 fire season by Richardson's Airway. In 1950, Ralph and his wife Gini started Richardson's Airway, which was a crop-dusting business and in 1951 they started Richardson Aviation, which was a flight school and fixed base operator. They flew from the Yakima Airport, Washington. From 1953 to 1994, Ralph Richardson operated an aerial crop spraying business. In September 1960, N9866C was sold to Red Dodge Inc., Anchorage, Alaska.

In 1960, Dale Newton and Dick Foy participated in their first fire season with the surplus B-25 s/n 44-28834.

Newton and Foy operated for that season under the name of Western Air Industries based at nearby Redding Airport, California. The following year they purchased their first two B-17 Flying Fortresses and they changed their name into Aero Union Corporation. During 1962, Aero Union moved its base to Chico Municipal Airport where it built up a major maintenance and engineering facility in the former U.S. Air Force hangars. Dale Newton was President; Dick Foy was Vice President and during the 1970-1980s another experienced tanker man Roy D. Reagan held a management position.

In the late 1960s, the company purchased a bulk of ten Fairchild C-119C Flying Boxcars from USAF disposals at Davis Monthan including a large stock of engines and spare parts stored in a hangar at Coolidge, Arizona.

In the 1970s, they bought a lot of Douglas C-54s and DC-6s. Aero Union Corp. became a large fire bomber operator and was a leader in introducing new aircraft types and improved retardant tank and delivery system designs. After years of controversies regarding the operation of the company and the safety of their planes, the USFS cancelled its contract in 2011 and the company was forced to shut down operations soon after. Aero Union Corp. also possessed a lot of warbirds like B-17s, Douglas A-26 Invaders and Grumman Avengers.

Colourful fire tanker N9865C of Western Air Industries later renamed Aero Union Corporation.

(Collection Richard Stratton via Ron Olsen)

Although a large company, it still had only two B-25s in service, N3695G and the B-25 from Western Air Industries, N9865C.

N3695G

This B-25 was a B-25J-15 with s/n 44-28926. After storage at Davis Monthan, she was purchased by John Bowman of Aerial Services, Anaheim, California in August 1959. She was registered as N3695G and was not a tanker. In May 1960, she was sold to Dale Newton. But in November 1961, this airplane was already sold to Marson Equipment & Salvage Co., Tucson, Arizona as part of the deal to purchase the two B-17s.

N9865C

The B-25J-15 with s/n 44-28834 was sold to Aero Spray Inc., Vancouver, Washington from the USAF storage in October 1958 and registered as N9865C. In May 1960, she was converted into fire tanker and in October 1960, she was sold to Western Air Industries, Redding, California. She flew with this company, later Aero Union, until 1963. She was overall aluminium finished with black engine nacelles. She flew as tanker number 30 with red/orange bordered black figures. In February 1963, she was sold to Clayton V. Curtis, Boise, Idaho where she was already sold a month later to E.D. Weiner at Los Angeles.

Below: This picture was taken at Falcon Field, Mesa around September 1968. She is parked in the hot sun at Mesa and was numbered 34C by Aviation Specialties Inc. The colours are already somewhat weathered, and this was the beginning of a long-term storage at the field. (Neil Aird)

Tanker #36 of Angel's Aerial Service of Pico-Rivera. This B-25 with registration N3512G was fitted with a large metal belly retardant tank. (napoleon130.tripod.com)

ANGEL'S AERIAL SERVICE, PICO-RIVERA, CALIFORNIA

One B-25 flew only for four months with a small firm called Angel's Aerial Service. This firm was located at Pico-Rivera California. This B-25J-25 had the USAAF s/n 44-30456. She was sold in July 1959 to Skyways System of Miami, Florida, and her registration of N3512G was reserved. The same month, she was sold to Hamilton Aircraft in Tucson, Arizona. By December of that same year, she was sold to Angel's Aerial Service of Pico-Rivera, California. She was modified by the addition of a borate tank in December 1959. She was sold in March 1960 to R. H. Hickish of Pico-Rivera, California. Thereafter, she had several owners and today she is a flying warbird in Soviet colours.

AVIATION SPECIALTIES INC., MESA, ARIZONA

The company Aircraft Specialties Inc. was founded around 1960 by Richard E. Packard as a tanker operator based at Mesa-Falcon Field, Arizona. In 1963, the company merged with Abe's Aerial Service of Safford. The new company was set up under the name of Aviation Specialties Inc. under the joint management of Richard Packard and Abe Sellards. They continued their spraying and fire-bombing operations. Their airplanes included the two B-25s of Abe's Aerial Service. During 1976, the spraying fleet was deployed on large-scale contracts for budworm forest spraying in Maine and Georgia. By 1977, some sections of the business were conducted under the name of associated company Globe Air Inc. Effective by April 1981, a business reorganisation resulted in all operations being transferred to Globe Air Inc. Sellards and Packard con-

tinued to operate the business until 1985 when the company was closed down. The airplanes were sold by an auction at Falcon Field in October 1985. The company had a lot of airplanes. During the 1960s large stocks of military disposals were acquired and rebuilt at Mesa as civil conversions. Among the airplanes were many helicopters, B-17s, PV-2 Harpoons, Douglas C54s, Lockheed Constellations and the two B-25s N6123C and N9117Z from Abe's Aerial Service Inc.

N6123C

One of the B-25s of Abe's Aerial Service Inc. was a B-25J-35, s/n 44-86893, and was registered in the U.S. as N6123C. Abe's Aerial Service was the aircraft's first civil owner. She was sold to Abe's in October 1958. In May 1959, she was fitted with tanks for agricultural spraying for her new job as a crop duster. In November 1965, she was sold to Aviation Specialties Inc., Mesa, Arizona and flew with number 34C. In 1969, she was withdrawn from use and stored derelict at Mesa until 1976. She was purchased by John Stokes of CenTex Aviation in San Marcos, Texas and flown to Chino, California for storage. In May 1977, she was purchased by Kansas City Warbirds of Kansas City, Missouri. Nowadays, she is part of the fleet of historic aircraft of the "The Flying Bulls" in Austria.

N9117Z

The other B-25 of Abe's Aerial Service was registered as *N9117Z*. This was a B-25J-20 with s/n *44-29199*. After storage at Davis Monthan, she was purchased by Abe's Aerial Service in January 1960. In April 1960, she was also converted into an air tanker and sprayer and flew as tanker #35C. In November 1965, she was sold to Aviation Specialties Inc., Mesa, Arizona. Like the other B-25 *N6123C* she was sold to John Stokes of CenTex Aviation in 1975. The airplane is currently operated by the National Museum of World War Two Aviation, Colorado Springs, Colorado and is flown as "In The Mood".

Biegert Brothers Inc. had three Mitchells for a very short time. This is N3155G after Biegert had sold her to Ontario Flight Service, Ontario, California. Here she is seen in her last days with Ontario Flight Service. (Collection Wim Nijenhuis)

Number 35C was overall aluminium finished with a white fuselage top and had like #34C also red accents and red engine nacelles. She is here parked at Falcon Field in 1972/1973.
(Jay Sherlock)

BIEGERT BROTHERS INC., LINCOLN, NEBRASKA / PHOENIX, ARIZONA

A company that had three B-25s for a short time was Biegert Brothers Inc. These were *N3155G, N3337G* and N9877C. The brothers Max L. and John Biegert established an agricultural dusting and spraying business in 1947. They operated under the names Biegert Bros. and Biegert Bros. Aerial Spraying, operating from Lincoln, Nebraska. In April 1953, the brothers moved into heavy spraying aircraft when they acquired a derelict B-17F. They rebuilt the airplane to airworthy condition as a sprayer, installing seven chemical tanks in the cabin as well as two additional underwing chemical tanks in fuel drop tanks from a USAF Lockheed

F-94 Starfire jet fighter. The B-17 started operating on a two-year contract for Central Aircraft Corp., Yakima, Washington for large scale mosquito spraying across the U.S.A. In 1956, a second B-17 was acquired and converted into a sprayer. By 1959, the company moved to Phoenix Arizona. Circa 1974, a new corporation was set up under the name of Biegert Aviation.

N3155G

This was a B-25J-25 with s/n *44-30832*. After storage at Davis Monthan, the airplane was sold to Aviation Rental Service, St. Paul, Minnesota in October 1958 and registered as

#44-29145 still in service with the USAAF. Given the new nose, this photo may have been taken shortly after the nose modification. Biegert bought the airplane in September 1961 and sold her already a month later to Crowl Dusters. (Collection Wim Nijenhuis)

N3155G. Already in January 1959, the B-25 was purchased by Max Biegert from Phoenix. Just three months later, Biegert sold her to Ontario Flight Service, Ontario, California where she was used for mapping and survey work. She was modified by installation of radar in her tail, additional seating and windows, a cargo floor, and an aerial mapping camera.

N3337G
This was a B-25J-30 with s/n 44-86891. The airplane was purchased from the USAF disposal sale by Max Biegert in April 1959. She was registered as N3337G. In December 1959 she was sold to L.K. Roser, Phoenix, Arizona and converted into an air tanker.

N9877C
The third B-25 of Biegert Brothers was a B-25J-20 with s/n 44-29145. She was in service with the USAAF after factory modification into an 8-gun B-25J. In 1944, she was initially assigned to the 3rd Army Air Service Command Replacement Pool where she was prepared for combat assignment. In November 1944, she was assigned to the 340th Bomb Group, 489th Bomb Squadron.

She was not flown on any combat missions but was used for unit evaluations of the gun nose modification. After the war, like N3155G mentioned earlier, this airplane was purchased from Aviation Rental Service, St. Paul. She was purchased in September 1961 and registered as N9877C. Biegert sold the airplane a month later to Crowl Dusters, Phoenix, Arizona.

BLUE MOUNTAIN AIR SERVICE / HILLCREST AIRCRAFT CO. INC., LA GRANDE, OREGON

Blue Mountain Air Service was an agricultural spraying company with many Boeing Stearman biplanes and some other airplanes. Also, between 1958 and 1962, it tanked and operated a number of USAF surplus B-25s for fire bombing. Blue Mountain Air Service was founded in the early 1950s by Eldon Down, manager of the La Grande Airport. In 1962, the company was

taken over by Gerald D. "Jerry" Wilson and renamed Hillcrest Aircraft Company. Wilson founded Wilson Aviation Industries Inc. at Orofino, Oregon in 1956 and now has been providing specialty aviation services to its customers for more than 60 years. At its beginning, Hillcrest owned and operated four Stearman biplanes for crop dusting. In the late 1950s, Hillcrest expanded its operations to include firefighting, using World War II aircraft that included B-25s, B-26s, Grumman TBMs and one B-24. Wilson established Hillcrest Aircraft Company in 1962 when he took over Blue Mountain Air Service Inc. The 1962 USFS ban on B-25 tankers because of structural failures during retardant drops no doubt influenced the merger of Blue Mountain Air Service into Jerry Wilson's new Hillcrest Aircraft Co., which would expand both agricultural and fire-bombing operations. An associated company was La Grand Air Inc., Oregon to which several aircraft were registered in the 1960s. After Hillcrest Aircraft moved its main operating base to Lewiston, Oregon, La Grande Air reformed as La Grande Air Service as a fixed base operator at La Grande Airport and in 1976 began fire bombing operations with a Douglas DC-7. Nowadays, Hillcrest's corporate headquarters are located in Lewiston, Idaho with a satellite base in Orofino, Idaho and specialised as rotary wing operator using the latest Bell helicopters.

Nine B-25 were owned by Blue Mountain Air Service and Hillcrest Aircraft Company respectively. These had the civil registration numbers N2887G, N2888G, N3503G, N3521G, N5239V, N8194H, N9642C, N9856C and N9857C. Seven of these B-25s were purchased from the USAF storage at Davis Monthan directly or via Ace Smelting Inc. Phoenix, Arizona.

N2887G
This was a B-25J-30 s/n 44-86716. This airplane was purchased from the USAF disposals at Davis Monthan in October 1958. She was registered as N2887G. In May 1959, she was converted into a fire tanker and fitted with a 1,000-gallon tank. She was sold to Idaho Aircraft Inc. at Boise, Idaho in March 1961.

Right:
One of the little-known pictures of a B-25 in Blue Mountain/Hillcrest colours. In December 1959, N3521G was purchased by Blue Mountain and in July 1962 she was transferred to Hillcrest Aircraft Co. The airplane was destroyed by fire on the ground after an emergency landing at Red Bluff, California on 6 May, 1967.

(Collection Wim Nijenhuis)

N2888G
B-25J-30 s/n 44-86872 has almost the same history with Blue Mountain. Purchased in October 1958 from the USAF disposals and registered as N2888G. In March 1961, she was also sold to Idaho Aircraft Inc.

N3503G
B-25J-25, s/n 44-30085, was purchased by Blue Mountain in July 1959 from USAF disposals sales at Davis Monthan via Ace Smelting Inc. She was registered as N3503G. She was converted into a fire tanker in July 1958 and fitted with a 1,000-gallon tank. Unfortunately, on 22 July, 1960 the airplane crashed and was destroyed while fire bombing near Baker, Oregon.

N3521G
In August 1959, B-25J-30 with s/n 44-86853 was purchased by Earl "Red" Dodge, Anchorage, Alaska from USAF storage. She was registered as N3521G. In December of that year, the airplane was purchased by Blue Mountain. She was converted into a fire tanker with a 1,000-gallon tank. In July 1962, she was transferred to Hillcrest Aircraft Co. In May 1963, she was converted for pesticide spraying with wing leading-edge spray bars and in July 1963, she was converted back for State agency firefighting operations. On 6 May, 1967 she was destroyed by fire on the ground after an emergency landing at Red Bluff, California.

This is the fuselage of N2888G at Robins AFB, Georgia in 1988. After this B-25 was sold to Idaho Aircraft Inc. respectively Dennis G. Smilanich, Boise, Idaho she was reported derelict and stripped at Boise. The fuselage was transported to Robins to be rebuilt for static display. (Michael Baldock)

N5239V

After open storage at Davis Monthan, the B-25J-25, s/n 44-30079, was sold to Mack Ballard, Riverside in September 1957 and registered as N5239V. She was converted into a fire tanker with a 1,000-gallon tank in July 1958. In January 1959, she was purchased by Blue Mountain and already sold again three months later to Ballard & associates, Riverside, California.

N8194H

B-25J-30, s/n 44-86749, was purchased by Blue Mountain in July 1959 from USAF disposals sales at Davis Monthan via Ace Smelting Inc. She was registered as N8194H. In July 1962, she was transferred to Hillcrest Aircraft Company, La Grande, Oregon. In February 1963, the airplane was sold to H.M. Trussell, Houston, Texas.

N9642C

This airplane was purchased by Blue Mountain by disposals from the Davis Monthan storage in June 1959. This was a B-25J-20 with s/n 44-29725. The B-25 was struck-off the United States Civil Register of Aircraft in September 1959.

N9856C

This was a B-25J-10 s/n 43-28204. She was bought in September 1958 from the USAF and then registered as N9856C. In May 1959, a 1,000-gallon retardant tank was added, and she was used as a fire bomber. Together with N2887G and N2888G she was sold to Idaho Aircraft Inc. in March 1961 where she flew as a tanker.

N9857C

The ninth B-25, s/n 43-28059, a J-5 model, was stored at Davis Monthan in 1957 and purchased by Blue Mountain in September 1958. She was registered as N9857C. She was converted into an air tanker in April 1959 and fitted with a 1,000-gallon tank. In December 1959, she was sold to Earl "Red" Dodge, Anchorage, Alaska where she flew as tanker #4.

Blue Mountain Air Service sold N9857C in December 1959 to Earl "Red" Dodge, Anchorage, Alaska where she flew as tanker #4. Here she is seen with the colours and nose art of Red Dodge.

(Collection Wim Nijenhuis)

CAL-NAT AIRWAYS, GRASS VALLEY, CALIFORNIA

Cal-Nat Airways was a Grass Valley firm that supplied pilots and borate tanker planes for forest firefighting. Cal-Nat Airways Inc. was founded in 1959 by Robert James Stevenson at Grass Valley, California. In 1958, pilot Stevenson and his Cal-Nat Airways began a 10-year association with the state and federal firefighting services and with Nevada County. Stevenson supplied planes, pilots, and airport management expertise on contract. He had purchased several World War II surplus Navy aircraft. Among them were

N5256V sitting at Grass Valley, California after she was withdrawn from use by Cal-Nat Airways in 1968. She was stripped of her engines. (Gordon Reid)

N9455Z in the 1960s stored at Grass Valley. She still has the red paint and tanker number 82 of her previous owner Les Bowman/Mantz. (Randall Grahek)

N3515G at Grass Valley was reported for sale in 1966 and probably scrapped.
(Collection Wim Nijenhuis)

Grumman F7F Tigercats, TBM Avengers and one Northrop P-61 Black Widow night fighter. Cal-Nat Airways quickly became a major USFS fire tanker player through the 1960s with a large and varied fleet including four B-25 Mitchells. In 1968, Bob Stevenson sold Cal-Nat Airways to Sis-Q Flying Service Inc. which continued flying the vintage WWII aircraft well into the 1970s. Sis-Q Flying Service was owned and operated by Robert Davis, a pioneer in the development of aerial firefighting aircraft.

The four B-25s were N5256V, N9455Z, N3515G and N2887G. The first three were all purchased from Les Bowman, Long Beach, California in December 1964. By 1968, all three were withdrawn from use, stripped, and stored at Grass Valley.

In 1970, N5256V was sold to Ralph M. Ponte at Grass Valley, N9455Z was sold to Sis-Q Flying Service and N3515G was reported for sale by U.S. Civil Register in 1966. Probably she was scrapped. The fourth B-25 from Cal-Nat Airways was N2887G, a B-25J-30 with s/n 44-86716. She was purchased from Loening Air at Boise, Idaho in April 1967. Just eight months later in December 1968, she was sold to DuPree Air Activities, Aguirre, Puerto Rico where she was broken up on a dump in 1982.

CHRISTLER AND AVERY AVIATION CO., GREYBULL, WYOMING

A major aerial spraying and firefighting company with four B-25s was Christler and Avery Aviation Co. at Greybull, Wyoming. Mel Christler and Morris Avery started Christler and Avery Aviation after World War II, offering aerial spraying services.

In 1949 they formed Christler and Avery Aviation in Greybull, Wyoming purchasing a surplus Douglas B-18 bomber and converting it into an agricultural sprayer. It was also retrofitted to haul ore from a uranium mine in the Big Horn Mountains.

In 1958, the company expanded its capabilities with the purchase of a Bell 47 helicopter. They used the helicopter in numerous challenging applications, such as mountain construction and the setting of power poles in rough terrain. Late in 1958, the company purchased four World War II-era PB4Y-2 Privateers from the U.S. Coast Guard. The large four-engine bombers were retrofitted for aerial firefighting, making Christler and Avery Aviation one of the early companies to contract with the U.S. Forest Service to use large aircraft to fight forest fires. They also purchased four USAF surplus B-25 Mitchells for spraying and fire bombing. In 1961, Mel Christler left Greybull and Morris Avery

purchased his share in the business and the company was reformed in August 1961 under the name Avery Aviation Inc., Greybull. Avery Aviation continued the same operations of fire bombing, pest spraying and fire ant baiting contracts, but increased the helicopter fleet for spraying and general charter. Aerial firefighting continued to be a major part of the company's business, but it also provided nonstandard aerial services, cloud seeding, mosquito and sagebrush spraying, and dropping dynamite on the Big Horn River to break up a massive ice jam.

When Morris Avery was suffering health problems, Privateer tanker pilot Gene Powers recruited experienced helicopter instructor Dan Hawkins from Texas to join the company to look after the rotary wing division. After the death of Avery in 1965, his wife, Reba, continued to operate the company with the assistance of employees Gene Powers and Dan Hawkins. She sold the company to them in 1969, and it continued to operate for decades as Hawkins & Powers. Hawkins and Powers Aviation's fire-bombing operations commenced with the Privateers taken over from Avery Aviation.

In the 1970s, the company purchased more airplanes and in the 1980s, they got access to more heavy tankers. Hawkins and Powers Aviation Inc. ceased business and closed in 2005 after the grounding of its PB4Y Privateer and C-130A Hercules fire bomber fleet due to tragic wing structural failures during fire attacks that summer. The flight operation site now serves as Museum of Flight

and Aerial Firefighting where numerous aircraft remain on display but no B-25.

Both companies Christler and Avery Aviation and Avery Aviation Inc. at Greybull had four B-25Js in service. They were all J-25 models and with the civil registration numbers N2849G, N3699G, N8195H and N9582Z.

N2849G

This was a B-25J-25, s/n 44-30077, stored at Davis Monthan in 1958. In 1960, she was sold to John M. Young, Beverly Hills, California and registered as N2849G. After having been owned for a short time by Dennis C. Moore, Boulder, Colorado, the airplane was sold to Christler and Avery Aviation at Greybull in September 1960. She was converted for agricultural spraying in March 1961. In August 1961, she was transferred to Avery Aviation Inc. In 1968, the Mitchell was sold to Filmways Inc., Hollywood, California and flew in the movie "Catch-22".

N3699G

In 1959, Fogle Aircraft, Tucson, Arizona was the first civil owner of the B-25J-25 with s/n 44-30801 from the storage at Davis Monthan. She was registered as N3699G. In January 1960, she was sold to Christler & Avery Aviation and transferred to Avery Aviation in August 1961. She flew as a sprayer and stayed with the company until 1968 when she also was sold to Filmways Inc. for use in the movie "Catch-22".

N8195H

The B-25 N8195H had s/n 44-30748. After the storage at Davis Monthan, she was sold to Ace Smelting Inc., Phoenix, Arizona, and the registration N3447G was reserved but was never used. She was purchased by Alson-Niblock Flying Inc., Elkhart, Indianapolis in May 1959 and registered as N8195H. In December 1959, she was sold to Christler and Avery Aviation and fitted with agricultural hopper and spray bars that were installed in January 1960. Like the other two

B-25 N2849G at Greybull in July 1966 with fire tanker belly retardant dump chute. (Geoff Goodall/Norm Avery)

Below: *N2849G on the ramp at Orange County Airport in late December 1968. She was purchased by Filmways in that year and she is here ready to leave for filming "Catch-22" in Mexico. She has already been painted in her basic film colours. (Phillips via AAHS)*

N3699G at Greybull in 1965, with fire and bait spreader installed under the fuselage. (Milo Peltzer)

N8195H derelict at Greybull in 1965. She was converted into a sprayer or duster and fitted with agricultural hopper and spray bars.

(Milo Peltzer)

Mitchells, she was transferred to Avery Aviation in August 1961 and sold to Filmways Inc. for the movie "Catch-22".

N9582Z

The fourth B-25 of the company was N9582Z, a B-25J-25 with s/n 44-30607. This airplane was sold from USAF storage to National Metals, Phoenix, Arizona in January 1960. She was registered as N9582Z. Christler and Avery Aviation bought the ship in September 1960. She was fitted with agricultural hopper and spray bars in February 1961 and they were removed in November of the same year. The airplane was transferred to Avery Aviation in August 1961. Finally, she left the company when she was sold to Desert Aviation Service, Phoenix, Arizona in September 1963.

1968, N8195H shown here in the camouflage paint scheme for the film "Catch-22" is ready for departure to Mexico. (Michael O'Leary)

CISCO AIRCRAFT, LANCASTER, CALIFORNIA

CISCO Aircraft Inc. was an early heavy aircraft tanker operator, providing agricultural and chemical spraying and fire-bombing services. CISCO stood for California Insecticide Service Company and was based in Lancaster, California. Owner was Zack C. Monroe. The company had a large number of Grumman TBM Avenger tankers and they were painted with large letters "CISCO" on the fuselage sides. It also had two B-25 tankers. The company ceased its operations by 1964.

Tanker #66, N3446G, photographed a few months before she crashed on 22 July, 1960 in the San Gabriel Mountains. (Milo Peltzer)

N5256V sitting at Grass Valley, California after she was withdrawn from use by Cal-Nat Airways in 1968. This ship was red with a white fuselage top and light grey bottom. On the rear fuselage she has still remnants of the name CISCO. (Collection Wim Nijenhuis)

COLCO AVIATION INC., ANCHORAGE, ALASKA

Colco Aviation Inc. was a tanker operator which provided fire bombing as well as a range of Alaskan services such as dropping sand on frozen rivers. From 1963, the company was stationed at Anchorage and moved to Fairbanks in 1969. Despite the USFS ban on B-25 fire bombers from 1962, the B-25 continued to be used in Alaska operating for State authorities. Colco Aviation had four B-25s: N9444Z, N9936Z, N9937Z and N88972.

N3446G
One B-25 was N3446G, a B-25J-30 with s/n 44-31466. She was purchased from John Bowman of Aerial Services at Anaheim, California in May 1959. She flew as tanker #66. On 22 July, 1960, this tanker impacted the earth during a water bombing run in Mill Canyon in the San Gabriel Mountains on the Magic Mountain Fire. The crew of three, James Armstrong, Charles Franco and John Bowman did not survive the crash. What remains at the site is a section of a wing, part of the stabilo, all three landing gear struts and both engines. Most of the rest of the aircraft melted away during the post-crash fire or was removed.

N5256V
The other B-25 of CISCO was N5256V. This ship was purchased from Les Bowman at Long Beach in February 1962. She was fitted with agricultural spray bars in June. In November 1963, the ship was repurchased by Les Bowman and later sold to Cal-Nat Airways.

N9444Z
The B-25J-25 with s/n 44-29943 was purchased by Colco from Kenneth Patterson, Tucson, Arizona in May 1961. The airplane was registered as N9444Z in 1959 when she was bought by National Metals, Phoenix. Early 1970s, she was retired and stripped for parts. By 1976, she was derelict at the airport fire service at Fairbanks.

N9444Z was sold as surplus in February 1959 and went through several owners. She ended up with Colco Aviation at Fairbanks, as seen in this picture from 1980. In later years, she was stored in the desert by Aero Trader. (Barry Greenfield)

drop on a fire that was in the Livengood/Rampart area of northern Alaska, the B-25 air tanker crashed due to a late pull up. Two crew members were killed.

N88972

A fourth B-25 of Colco was N88972, a B-25D-30. After she had flown for some years in Canada with a Canadian civil registration, she was back in the U.S.A. and was registered as N88972. She was purchased by North Star Aviation Corp. at Fairbanks in December 1966. A river sanding/fire retardant tank was installed in August 1967. In April 1969, the ship was sold to Colco Aviation. By 1977, she was retired as a fire bomber and was stored at Fairbanks. In May 1980, the ship was sold to Noel Merrill Wien, Anchorage. Later in the 1980s, the airplane went to The Fighter Collection at Duxford in the U.K. and flew as "Grumpy".

N9936Z

This B-25J-25 s/n 44-30756, was stored at Davis Monthan in December 1958. In July 1959, she was purchased by National Metals, Phoenix and registered as N9936Z. She was sold to Paul B. Hanson, Pacific Palisades, California in March 1961 and then to Colco Aviation Inc. at Anchorage in May 1961. There she flew as tanker #3. By 1975, she was retired, stripped and derelict at Fairbanks. In January 1981, she went to the Alaskan Historical Aircraft Society, Anchorage.

N9937Z

A similar history applies to N9937Z. This B-25J-1, s/n 43-3910, was also purchased from Paul Hanson in May 1961. She also was converted into a tanker and had #2. Unfortunately, on 27 June, 1967, during a retardant

The overall natural aluminium finished N9936Z was tanker number 3 of Colco Aviation Inc., here photographed at Fairbanks in September 1968. The engine nacelles were black with a blue cowling ring. (Neil Aird)

N9937Z at Merrill Field, Alaska in 1962. At right, in September 1966, and now numbered 2 on her vertical tails. This airplane crashed on 27 June, 1967, at Rampart. (Collection Wim Nijenhuis, Neil Aird)

N88972 stored at Fairbanks in 1979. She still has the company name of her previous owner North Star Aviation on her nose.

Below: *Two years later at Fairbanks in September 1981, then owned by Merrill Wien.*

(Geoff Goodall)

CROWL DUSTERS, PHOENIX, ARIZONA

Crowl Dusters Inc. was an agricultural operator. In the early 1950s, the company was founded by Clifford R. Crowl from Glendale, Arizona. The company was based at Air Haven Airport, Phoenix and had some surplus Navy airplanes. For firebombing they had for about four years a B-25 that was purchased from Max Biegert at Phoenix in October 1961. This was a B-25J-20 and registered as N9877C. She had tanker number C39. The airplane was sold in September 1965 to Allied Aircraft Sales at Phoenix. Finally, the airplane ended in Mexico. See the chapter Mexico.

Tanker C39 was used for fire bombing for about four years by Crowl Dusters. At right, a 1/48 scale model kit of the airplane made by the author gives a good impression of the colours (Collection Wim Nijenhuis)

DONAIRE INC., PHOENIX, ARIZONA

Donaire Inc. operated from Deer Valley airfield near Phoenix and in the early 1960s, they moved to Buckeye airfield, Arizona. This aerial application business was founded in 1959 by W. Donald Underwood. An associated company was Donaire Sales Inc. The company had some warbirds, especially A-26 Invaders and four B-25 Mitchells. The tanker operations of Donaire wound down in the late 1960s. The four Mitchells were N9079Z, N9080Z, N9452Z and N9655C.

N9079Z
Was sold in December 1959 to Donair Sales at Phoenix from the USAF disposals at Davis Monthan. This was a B-25J-25 with s/n 44-30734. In March 1960, a spray tank was installed. Her registration was incorrectly assigned as N9080Z and was corrected to N9079Z in April 1961. She was operated under a Forest Service contract as tanker #32 until 1962. She sat idle for four years until she was sold in March 1966 to Robert Clements of Indiantown, Florida. In January 1968, she was sold again to Clements and Howe Aviation, Indiantown.

N9080Z
This plane was also sold in December 1959 to Donair Sales at Phoenix from the USAF disposals at Davis Monthan. This ship was a B-25J-10 with s/n 43-36074. In March 1960, she was converted into an air tanker with a 1,150-gallon tank. She was tanker #04 and later #C04. Mid-1960s, she was sold to Mexico and registered as XB-NAJ.

N9452Z
This B-25 was purchased in August 1965 from Sprung Aviation, Tucson, Arizona. The B-25J-25 with USAF serial 44-30649, had no civil conversion and retained her in the United States Air Force scheme. In June 1968, she was sold to Tallmantz Aviation, Orange County for use in the film "Catch-22".

N9655C
The fourth B-25 was a J-25 model and purchased from J.E. Gardner, Phoenix in November 1959. The B-25 had s/n 44-30159 and was registered as N9655C. She was converted into an air tanker in April 1960 and fitted with a 1,150-gallon retardant tank. In September 1964, she was sold to Universal Air Sprayers, El Reno, Oklahoma. In 1974, she was sold again to Crab Orchard, Tennessee and finally removed from the U.S. Civil Register in October 1975.

During the early 1960's, N9079Z was highly modified into a tanker to fight forest fires. She flew with Donaire Inc. as tanker #32. (Delaware Aviation Museum)

DOTHAN AVIATION CORP., DOTHAN, ALABAMA

In the 1960s and 1970s, a man by the name of Hugh Wheelless operated a very busy aerial spraying operation out of Dothan, Alabama. Dothan Aviation Corporation was a corporation engaged in, among other things, the business of crop dusting. It owned and operated several airplanes which it used for crop dusting and employed a number of commercial pilots to operate these dusting machines. Dothan began in the early 1950s as a single-engine aircraft agricultural business. Hugh W. Wheelless founded Dothan and began commercial services with charter flights and crop dusting. An associated company was Dothan Leasing and Rental Co. Inc. The company was based at Dothan Municipal Airport before moving to its own grass field at Wheelless Airport on the rural outskirts of Dothan. The office of Dothan was a building on the northeast corner of Wheelless Airport. From 1959, they specialised in wide-area aerial application operations using heavy aircraft, bidding for Federal and State pest eradication and forest spraying programs. After Hugh Wheelless died in February 1970, the business was managed by his son Hugh W. Wheelless Jr. Dothan Aviation Chief Pilot was Virgil Fenn, who logged 3,000 hours on the B-17s alone. He was also Vice President of Dothan Aviation Corporation. Dothan had several PV-2s, B-17s, Martin 404s and a fleet of Stearmans, Grumman Ag-Cats, and Piper Cub dusters and sprayers. Dothan Aviation ceased operations in 1978. Hugh Wheelless Jr. was still president and was personally involved in the disposal of assets until at least 1983. The Wheelless family closed the airport in 1992 and the land became a housing development.

Apart from the named aircraft, Dothan flew also with six B-25s. These were N3525G, N9077Z, N9443Z, N9462Z, N9463Z and N9552Z.

N9077Z was converted for agricultural spraying or dusting. She was purchased by Dothan in February 1960. She is seen here doing some testing of spreading equipment in 1959 and at Wheelless in 1974.

(Richard Goldman, Collection Gary Lewis)

N3525G

This plane was stored at Davis Monthan and sold to Leo Wendler of Wendler Air Ag, Dighton, Kansas City in July 1959. She was registered as N3525G. In June 1962, the airplane was purchased by Hugh Wheelless of Dothan Aviation Corp. she was struck off the civil register in June 1973. This B-25 was a J-30 model with s/n 44-86694.

N9077Z

A B-25J-5 with s/n 43-27868 was purchased by Dothan in February 1960. The airplane was registered as N9077Z and purchased from Fogle Aircraft Co, Tucson, Arizona. She was converted for agricultural spraying or dusting in January 1962. In October 1975, the B-25 was sold to John Stokes, San Marcos, Texas. Nowadays she flies as "Yellow Rose" of the Commemorative Air Force at San Marcos.

N9443Z

This ship was a B-25J-15 with s/n 44-28765 and registered as N9443Z. She was sold to Carma Manufacturing Co., Torrance, Califor-

nia in April 1960 and to Dothan Aviation in September 1961. After a period of nearly 15 years, she was sold to John J. Stokes at San Marcos in October 1975.

Below: N9443Z in the 1960s. The airplane was overall aluminium finished with a white fuselage top and red tips on top and bottom of the vertical tails. (SDASM)

N9462Z

Another B-25 sprayer of Dothan was s/n 44-30535, a B-25J-25. In August 1960, she was bought from National Metals, Phoenix. The plane was registered as N9462Z. She was fitted with an agricultural spray tank and bars in April 1961. After many years of service, she was sold to Vicki Meller at Burbank, California in July 1975. The plane later flew as warbird "Iron Laiden Maiden".

N9463Z

This B-25 was also purchased by Dothan from National Metals in August 1960. This was s/n 44-31004, a B-25J-30. She was also converted into a sprayer. In February 1974, she was donated to USS Alabama Battleship Memorial Park, Mobile, Alabama.

Wheelless Airport in April 1975. N9443Z has been retired and her former USAF markings could be seen through the faded paintwork. (John Hevesi)

N9462Z at Dothan in July 1973. She was fitted with an agricultural spray tank and bars. She still has some USAF markings. (Bob Laeder)

N9463Z at Dothan, she was also converted into a sprayer and here photographed in July 1973. (Bob Laeder)

B-25J-10 with s/n 43-35972 was registered as N9552Z. In December 1960, she was fitted with an agricultural tank and spray bars. This airplane served many years with Dothan Aviation Corp. (Collection Dick Phillips)

N9552Z

Finally, Dothan had the B-25J-10 with s/n 43-35972. This airplane too, was purchased from National Metals in August 1960. She was registered as N9552Z. In December 1960, an agricultural tank and spray bars were fitted. This airplane also served many years with Dothan and the company sold the ship in August 1975 to John Stokes at San Marcos. Nowadays, she is still a flying warbird named "Maid In The Shade".

FRONTIER FLYING SERVICE, FAIRBANKS, ALASKA

Retired Air Force Col. Richard McIntyre founded Frontier Flying Service in 1950, catering to Alaska bush communities, primarily around Fairbanks. The company provided charters throughout the state as well as mail service for Wien Air Alaska. In 1974, Frontier was purchased by John Hajdukovich. The company continued to grow with its acquisition in 2005 of Cape Smythe Air Services, taking on that company's equipment and infrastructure. In 2008, the owners of Frontier Flying Service (John Hajdukovich) and the owners of Hageland Aviation (Mike Hageland and Jim Tweto) agreed to form a new parent company (HoTH). This action created the beginnings of an "air group" allowing the parent company to make acquisitions. Frontier now operates under the brand Ravn Connect.

In the 1960s, Frontier Flying Service operated for nearly one year a B-25 Mitchell. This was a B-25J-30 with s/n 44-86791. This airplane was flown to storage by December 1957 and purchased by Ace Smelting, Phoenix, Arizona in May 1959. A civil registration of N8196H was assigned. She was then sold to Merrill and Richard Wien at Fairbanks, Alaska. In June 1961, she was sold to Merric Inc. of Fairbanks, were she was flown as tanker #5 in a red and white colour scheme. She was sold to Frontier Flying Service of Fairbanks, Alaska in May 1962. In February 1963, she was sold to Aero Retardant of Fairbanks. After she was sold to a few other owners in Alaska, the airplane was sold to Aero Heritage Inc. of Melbourne and ferried to Australia. See chapter Australia.

In June 1961, N8196H was sold to Merric Inc. at Fairbanks where she flew as tanker #5. She was sold to Frontier Flying Service of Fairbanks, Alaska in May 1962. It is not known if she flew with Frontier in the same red and white colours. (Geoff Goodall)

IDAHO AIRCRAFT COMPANY, BOISE, IDAHO

The Idaho Aircraft Company was incorporated on 12 February, 1959 in Boise, Idaho by Milton and Dennis Smilanich. By 1961, the company was a fire-bombing contractor to USFS and Bureau of Land Management. About ten years later, it withdrew from air tanker operations, but the company continued as a general aviation business at Boise Airport into the 1990s. The company owned four B-25s, N2887G, N2888G and N9856C and for a short time N9116Z.

N2887G

This airplane was a B-25J-30 with s/n 44-86716. She was purchased from the USAF disposals at Davis Monthan in October 1958 and was registered as N2887G. She was converted into a fire tanker and fitted with a 1,000-gallon tank by Blue Mountain

Not the best, but a rare picture showing an Idaho Aircraft Company B-25 and TBM at work dropping fire retardant in Idaho. This picture was published in "Rudder Flutter", Official Publication of the Idaho State Department of Aeronautics, August 1961, Vol 17 (4):1-6. (Geoff Goodall)

Air Service. In March 1961, she was sold to Idaho Aircraft Inc. In May 1962, she was transferred to Dennis G. Smilanich at Boise. He sold the B-25 to Loening Air Inc. at Boise in June 1965. The airplane was finally broken up on a dump at Salinas, Puerto Rico in 1982.

N2888G

This airplane was also a B-25J-30 and had s/n 44-86872. She was stored at Davis Monthan in December 1957, sold to Blue Mountain Air Service in October 1958 and registered as N2888G. In March 1961, the airplane was sold to Idaho Aircraft Inc. and in May 1962, transferred to Dennis G. Smilanich at Boise. She was stripped for parts and reported derelict and stripped at Boise by 1968. In 1983, she was acquired by the Pacific Museum of Flight. Later she was displayed at the United States Air Force Museum at Robins AFB, Georgia and currently as "The Little King".

N9856C

This airplane was a B-25J-10 and had s/n 43-28204. She was bought in September 1958 from the USAF and then registered as N9856C. She was modified to a fire bomber and together with N2887G and N2888G, she was sold to Idaho Aircraft Inc. in March 1961 where she flew as a tanker. In May 1962, she was also transferred to Dennis G. Smilanch. He sold the airplane to Filmways Inc. at Hollywood in September 1968 where she flew in the movie "Catch-22".

N9116Z

This B-25J-20 with s/n 44-29287 was stored at Davis Monthan in 1958. Idaho Aircraft Inc. purchased the airplane in 1963 and converted her to a tanker. But already in the same year, she was sold to Arthur R. Siegel at Miami, Florida. Three years later in 1966, she was exported to Bolivia and registered as CP-796.

N9116Z photographed in 1962 at Boeing Field, Seattle, Washington just before she was purchased by Idaho Aircraft Inc.
(David J. Gauthier)

N2888G at Boise. She was stripped for parts and reported derelict at Boise by 1968.
(Collection Wim Nijenhuis)

N9856C was owned by Dennis G. Smilanch until September 1968. Then she was sold to Filmways Inc. for filming "Catch-22". Here is the bomber at Burbank in late 1968, shortly before departing for Mexico where the film recordings were made. (Milo Peltzer)

JOHNSON FLYING SERVICE, MISSOULA, MONTANA

Johnson Flying Service owned N9088Z until December 1966. Here she is seen in 1968, then owned by Ed Thorsrud and flew as tanker #8Z, based in Alaska. (Collection Wim Nijenhuis)

Johnson's Flying Service of Missoula, Montana, maintained Forest Service contracts well into the 1960s and used both Ford Tri Motors and Travel Airs on floats or wheels to haul smoke jumpers and freight to remote regions. Although better known for its bush flying and forays into the regional airliner business, Johnson Flying Service also operated a small fleet of P-2H Neptune's in the air tanker role as well as a B-25. Johnson Flying Service provided reliable service during the infancy of aerial detection, cargo drops, and smoke jumping for three decades.

Johnson Flying Service was incorporated at Missoula on 6 December, 1928 by Bob Johnson. Passenger air charter and contract service flights began late April 1929 and U.S. Forest Service freighting contracts begun in May 1931. The company built up a specialist contractor role with the USFS and other Government agencies, providing transport and forest fire control services to inhospitable Tall Timber regions of the Northern Rocky Mountains. This included aerial seeding and dropping parachuting "smoke jumper" fire fighters and equipment. Mainstays of the company were Ford Trimotors. After the Second World War, Johnson standardised on Douglas C-47s, Curtiss C-46s, Beech 18s and the survivors of the pre-war Ford Trimotors and Travel Air's.

A new subsidiary, Johnson Airlines, was formed in January 1954 to offer DC-3 flights between Seattle and Spokane and to the tri-cities of Pasco, Richland, and Kennewick. In the mid-1950s, USFS requested Johnson Flying Services to investigate aerial fire bombing of forest fires. The first three Grumman TBMs were acquired from Navy surplus and where fitted with retardant tanks constructed by the company. Supplemental air carrier work was expanded with the purchase of a DC-4 and Lockheed Electra's operating under names Johnson Air and later Johnson International Airlines. These large transports were used to carry USFS personnel during the summer months, and general charter for the rest of the year. The workforce in 1964 exceeded 50, although some pilots continued to be employed on a seasonal basis to spray weeds or fight fires. Johnson Flying Services and subsidiary companies were dissolved and disincorporated effective in January 1977.

The B-25 operated by Johnson Flying Service was N9088Z. After her USAF service, she was flown to storage in August 1958. In December 1959, she was sold surplus to National Metals from Phoenix, Arizona and registered as N9088Z. She was immediately sold to Johnson Flying Service. The B-25 was converted for the use as a fire suppression aircraft by installing a chemical tank in the bomb bay. In December 1966, she was sold to Edgar Thorsrud at Missoula and based in Alaska as tanker #8Z.

PARSONS AIRPARK INC., CARPINTERIA, CALIFORNIA

Louis D.W. Parsons operated his firm from Parsons Airpark. This was a small private airport he constructed in Carpinteria, California around 1946-1947. The Airpark was evidently closed at some point around 1966 and 1967. Parsons piloted single-engine airplanes as well as helicopters. He also flew bombers, converted into tankers, to fight Southern California wildfires. In the 1950s, movie stars would sometimes fly in on their way up to Santa Barbara. Louis Parsons has been associated with eight companies, according to public records. The companies were formed over a fifty-one-year period with the most recent being incorporated in August 2007. Parsons Airpark Inc. was his first company incorporated in 1955.

The cover of the journal Fire Engineering from April 1959 shows a picture of Parsons' tanker #91 at work. (

Collections Don Croucher)

N10564 during flight test of air tanker firefighting for the USFS in 1962. On the rear fuselage the text "U.S. Dept. of Agriculture USFS" was applied. The B-25 was overall natural aluminium finished with black engine nacelles and day-glow sections on the nose, wings, and vertical tails. Note the research equipment on the fuselage top and in the nose. (AFFTC/HO)

N10564

One of Parsons' warbird fire fighters was the B-25J-15 with s/n 44-29887. This Mitchell was modified by Hayes in 1953-1954. After Hayes finished working on the plane, the USAF reassigned her to Wright-Patterson AFB and then to Eglin AFB, Florida. In November 1957, the government declared the Mitchell obsolete and sold her to Les Bowman Engineering at Long Beach, California. She was registered as N10564. The next month she was purchased by Louis Parsons, owner of Parsons Airpark. He contracted with AiResearch Aviation in Los Angeles to modify N10564 to conduct fire bombing missions. In August 1958, AiResearch installed tanks holding 1,200 gallons of the firefighting chemical borate and Parsons flew this "borate bomber" for several fire seasons. Parsons flew her as tanker #91. In late July 1960, four B-25 bombers crashed while dropping borate on forest fires near Magic Mountain in southern California. Two crewmen died in each crash. This string of tragedies led the USFS to ban the B-25 from fire bombing operations. Companies fighting fires with the B-25 protested and the USFS decided to study the type more closely. They selected N10564 for a series of flight tests conducted at Edwards AFB, California, in February 1962. The results confirmed the initial decision and B-25 firefighting operations never resumed. Parsons sold the B-25

Tanker #91 of Parsons. The name of the company is painted on the nose. (napoleon130.tripod.com, Sam Wood)

to Hemet Valley Flying Service in Hemet Valley, California, in May 1965. When she was purchased by Hemet Valley Flying Service, she was ferried to Hemet where her engines were removed for Super PBY Cat conversions. Hemet owned the B-25 for circa three years. She was stored in the open air until she was sold to Tallmantz Aviation, Santa Ana in October 1968. She flew in the film "Catch-22" as "Luscious Lulu".

N7687C
A second B-25 used by Parsons was N7687C, a B-25J-15 with s/n 44-28925. Declared surplus in January 1958, Parsons Airpark purchased her in April 1958. Her registration was assigned as N7687C in July 1958 after conversion into an agricultural sprayer. In July and December 1960, extensive modifications were made with the installation of electronic test equipment for Raytheon Corp. By August 1964, she was sold to Aerial Applicators Inc. and operated by Trans-West Air Service. Nearly four years later in May 1968, she was purchased by Tallmantz Aviation of Santa Ana, California for use in the film "Catch-22".

The second B-25 of Parsons was s/n 44-28925. She is seen here in her post-war USAF colours before she was purchased by Parsons in April 1958 and got the civil registration N7687C. (Collection Wim Nijenhuis)

PAUL MANTZ AIR SERVICES, SANTA ANA, CALIFORNIA

Paul Mantz was one of the founding fathers of aerial firefighting. His aerial activities ranged from cinema to transport and instruction. His firms United Air Services and Paul Mantz Air Services became more spe-

Paul Mantz (centre with cigar) at work. A camera is attached at the bottom of the nose of a B-25. Most likely this is NX1203.
(Life)

cialised in Hollywood VIP discrete transport. In July 1930, he set a world record of 46 outside loops flying a Fleet 2 biplane, and soon started flying for films. He was also Amelia

Earhart's advisor on her long-range raids. Before the war, Paul Mantz started his own flying and charter business, named United Air Services, at the United Air Terminal at Burbank, California. Motion picture flying, charter flying, forest patrol work and air ambulance flights filled Mantz' days. His charter service was popular with the Hollywood crowd, and his legendary Lockheed Vega called "Honeymoon Express" was in almost continual use by stars heading for Las Vegas marriage chapels.

After the war, he renamed his company Paul Mantz Air Services. The company handled tasks including flying daily film rushes into Hollywood from distant locations. In 1946, he paid the Reconstruction Finance Corporation $55,000.00 for what amounted to a whole airport full of airplanes. Most of the airplanes were located at Stillwater, Oklahoma. At the time, it was the world's sixth largest air force with 475 bombers and fighters. Among all these airplanes were ten B-25s, including s/n 43-4643. Mantz estimated that the fuel contained in the planes alone was worth more than he paid for the lot. The problem was that he had to get them off the field by a certain deadline. This turned out to be impossible. So Mantz ended up bringing dozen of the best ships to Burbank

The entrance of Paul Mantz Air Services in the 1950s at the Lockheed Air Terminal, Burbank. The air charter company was owned and run by movie stunt pilot, Paul Mantz.

(Los Angeles Public Library)

and sold the rest for scrap. Mantz stripped down three of the airplanes, two Mustangs and a B-25 bomber, and rebuilt them. Mantz would use the B-25 #43-4643, with the civil registration NX1203, as this personal aerial camera ship. This was to become his famous Cinerama photo plane. It was Mantz's favourite airplane that was used for film work. He called this airplane "The Smasher". Mantz and his B-25 were called in to film numerous other features in the 1950s. In 1956, Mantz moved his operations from Burbank to Orange County Airport, Santa Ana, California. He obtained a large section of the southeast portion of the airport to base his fleet of airplanes on. In 1961, Mantz merged his flight operations with Frank Tallman, retaining the Orange County base and incorporating Tallman's extensive aviation collection. Tallman, sixteen years his junior, took a lead role, becoming President of the new corporation. Mantz's Hollywood connections were invaluable, and he continued to play a behind-the-scenes role even though he took on fewer and fewer actual filming roles.

LES BOWMAN ENGINEERING, LONG BEACH, CALIFORNIA

In the 1960s, Mantz worked together with another company on firefighting. This was Les Bowman Engineering at Long Beach, California, a fire tanker operator in the same area. Les Bowman was an early fire tanker operator using the USAF disposals B-25 Mitchells. Bowman operated in partnership with Paul Mantz Air Services to promote aerial fire bombing. The enterprise of Bowman and Mantz was named Les Bowman and Paul Mantz Air Service and was a tanker operator that mainly used the B-25. In 1960, two B-25 tankers were sent to Caracas, Venezuela for the 1960 fire season. One was sold to the Venezuelan Government while the other B-25 returned to California. The following year, another B-25 tanker was ferried to Caracas for the 1961 season before returning to California.

In 1957, Les Bowman would buy two B-25s. In June 1953, these two-gun nosed B-25s were advertised for sale in full military configuration, including the guns. The planes were a B-25H-10 with s/n 43-4999 and a B-25J-10 with s/n 43-36075. Both had served for a short time in the Dominican Air Force. They were purchased by the Babb Company, Newark, New Jersey and registered as N3970C and N3969C, respectively. Then in 1957, they would be sold to Les Bowman, but the sale was not completed. The airplanes stayed derelict at West Trenton-Mercer County, New Jersey. In 1970, N3970C was donated to the Bradley Air Museum. At that point, she was disassembled and trucked to Bradley, Connecticut. It was there that she was restored to static condition. N3969C was broken-up and scrapped in 1969.

The B-25s N3970C (left) and N3969C (right) withdrawn from use and derelict at Mercer County Airport, West Trenton, New Jersey. Both would be bought by Les Bowman, but the sale was not completed. Finally, N3970C was restored and N3969C was scrapped.

(Collection Gary Fitton)

Paul Mantz obtained the B-25H 43-4643 in 1946. She was registered as NX1203. This picture shows the airplane at Phoenix Airport in 1947 with the nose markings of Weath-Air Inc., another company of Paul Mantz. At right, Paul Mantz in the cockpit in 1948. This ship could make rain or snow with certain water, dry ice and chemical dispensers. (William T. Larkins, Los Angeles Public Library)

Both companies of Mantz and Les Bowman operated B-25s. Six B-25s were operated by the Mantz company. These were NX1203, N3515G, N3516G, N7493C, N9455Z and N9456Z. Les Bowman operated eight B-25s with the registration numbers N3515G, N3516G, N5256V, N7493C, N7707C, N9455Z, N9456Z and N10564.

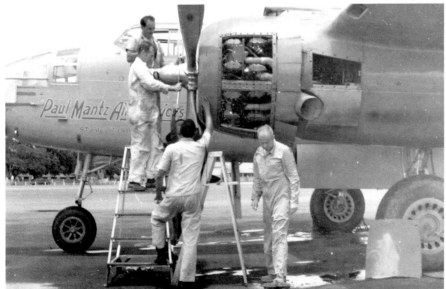

NX1203 in February 1953. She was then used for film work and the company name was replaced with Paul Mantz Air Services. (SDASM)

NX1203

Paul Mantz had for some time a B-25H-5 with s/n 43-4643. This airplane was declared surplus in October 1945 and stored at Searcy Field, Stillwater, Oklahoma. She was purchased by Paul Mantz in February 1946 for future film use. In May 1946, she was registered as NX1203 and used directly for some early film work. But before her career as a real camera ship, the B-25 was used as a rain maker. In June 1948, a new weather making company named Weath-Air Inc. had been established by Paul Mantz. Two surplus planes, this B-25H and a Grumman TBF Avenger were equipped to scatter silver iodide, dry ice and charcoal over various cloud formations to make both rain and snow. Cloud seeding was a relatively new business after the war. Paul Mantz created Weath-Air Inc. working on experimental devices with governmental agencies. NX1203 was used for some time for this new business. In the 1950s, the ship was modified as a camera platform and in 1956, she was transferred to Paul Mantz Air Services and in November 1961 she was sold to Tallmantz Aviation in Santa and registered as N1203. See the chapter Tallmantz.

N3515G

A sprayer from Les Bowman and Paul Mantz was the B-25J-30 with s/n 44-31042. She also was purchased from the USAF storage at Davis Monthan in July 1959 by National Metals Inc. She was registered as N3515G. In November 1959, she was converted into a fire tanker with a 1,000-gallon tank. She was sold to Zack C. Monroe, Burbank, California in February 1962 and flew as a sprayer. But the airplane was repurchased by Les Bowman in November 1963 and sold again one year later in November 1964 to Cal-Nat Airways at Grass Valley. She was reported for sale by U.S. Civil Register in 1966. Probably, she was scrapped

N3515G in the early 1960s. The fire tanker of Bowman and Mantz was sold to Zack C. Monroe, Burbank and later to Cal-Nat Airways at Grass Valley.
(Collection Wim Nijenhuis)

N3516G

This was a B-25J-15 with s/n 44-29035 and stored at Davis Monthan in 1957. In July 1959, she was sold to Les Bowman & Paul Mantz via National Metals Inc., Tucson, Arizona and registered as N3516G. In November of that year, she was converted into a fire tanker and fitted with a 1,000-gallon tank. She was operated by Paul Mantz Air Services as tanker #82. In March 1960, she was flown to Caracas, Venezuela for a fire tanker contract. In May 1960, she was sold to the Venezuelan Air Force.

N5256V

A B-25J-10 with s/n 43-28222 that served for a longer period with Les Bowman and Paul Mantz. The airplane was stored at Davis Monthan and purchased by Les Bowman in January 1958. She was registered as N5256V. In April 1959, the airplane was converted to air tanker and fitted with a 1,000-gallon tank. She was operated by Paul Mantz Air Services and flew as tanker #81. In March 1960, she flew to Caracas, Venezuela for a fire-bombing contract. In February 1962, she was sold to Zack C. Monroe of CISCO Aircraft Inc., Burbank, California, and the B-25 was fitted with agricultural spray bars. In November 1963, the ship was repurchased by Les Bowman. Finally, one year later she was sold to Cal-Nat Airways, Grass Valley, California. She was withdrawn from use by 1968

N5256V just after her Air Force days. The U.S. Air Force markings are still visible on her nose. (Collection Wim Nijenhuis)

A colourful N5256V as tanker #81 of Paul Mantz Air Services. (Dusty Carter)

N7493C

This was a B-25D-30 with s/n 43-3376. This was one of the early B-25 models which were converted into a fire tanker. The airplane was purchased by Les Bowman Engineering Co. in September 1957 and registered as N7493C. In October, she was sold to Paul Mantz Air Services and converted into a fire tanker. In 1961, she was based outside of Caracas, Venezuela and removed from the U.S. Civil Registry as destroyed in June 1970.

N7707C

After storage at Davis Monthan, she was sold to Les Bowman in June 1958. This B-25J-25 with s/n 44-30690 was registered as N7707C. Already in October of that year, she was sold to Jim Wells, Bruce Worthy, Gilbert Gann, Huntington Beach, California.

in August 1960 and fitted with a 1,000-gallon tank. She flew as tanker #82, the second B-25 with this number. In 1961, she was used as a tanker and based in Caracas, Venezuela. In February 1962, the airplane was sold to Zack C. Monroe, Burbank, California. She was repurchased in November 1963. After one year, she was sold to Cal-Nat Airways at Grass Valley in December 1964. After she had some other owners, she went to the United Kingdom and flew as "Big Bad Bonnie" in the film "Hanover Street". See the chapter United Kingdom/Hanover Street.

to Tallmantz Aviation, Orange County, California. In March 1962, terrain avoidance radar and a precision radar altimeter were installed for an FAA test program. She was used in the filming of "Catch-22" in 1969. She was sold to Donald Buchele of Columbus Station, Ohio in January 1971. In August of 1978, she was sold to F. Gene Fisher of Boiling Springs, Pennsylvania and in 1981, she was donated to the Mid-Atlantic Air Museum and restored.

N10564

This was a B-25J-20, s/n 44-29887. In November 1957, after the Mitchell was declared obsolete, she was sold to Les Bowman Engineering for $2,777. Bowman registered the bomber as N10564. But after a month, the bomber was sold to Louis Parsons, owner of Parsons Airpark in Carpinteria, California. In 1958, she was modified to conduct fire bombing missions.

Another colourful tanker was N9455Z #82 of Paul Mantz Air Services in 1960. This airplane replaced the first tanker #82 (N3516G) which had been left in Venezuela after a firefighting contract had been completed. (Carmen Zone)

N9456Z was purchased by Les Bowman and Paul Mantz in 1960, becoming part of the Paul Mantz Air Services fleet. The bomber transferred to Tallmantz Aviation in November 1961 when the company was founded and was primarily used as a contracted testbed for several years. She has red rudders, nacelles, and wing tips over a white background. (Milo Peltzer)

N9455Z

This Mitchell was a B-25J-25 with s/n 44-30210. The airplane was delivered to the US-AAF in December 1944 and stored at Davis-Monthan in December 1959. She was sold to National Metals Inc. in January 1960 and registered as N9455Z. In May 1960, she was sold to Les Bowman and Paul Mantz Air Services. She was converted into a fire tanker

N9456Z

The B-25J-25 with s/n 44-29939 was bought by Bowman and Mantz in May 1960 from the USAF Davis Monthan storage via National Metals Inc. She was registered as N9456Z. In November of 1960, she was sold to Paul Mantz Air Services where she would be used as a freight support for many movies. In November 1961, she was transferred

N3476G at Whitehorse, Yukon in September 1980. This was tanker #6 from Red Dodge and had wonderful nose art depicting a horse pulling an old-fashioned fire wagon. At the time, the ship was owned by Robert Schlaefli, Port Orchard, Washington. (Gary Vincent)

One of the companies that had several B-25s was Red Dodge Aviation Inc. at Anchorage, Alaska. Earl "Red" Dodge was an Alaskan charter operator, operating under the names Red Dodge Aviation Inc. and Red Dodge Inc. The company was engaged in the business of carrying commercial air freight between Anchorage, Fairbanks, and the North Slope of Alaska pursuant to the rights granted to it by the Alaska Transportation Commission. In the late 1960s, the company entered a partnership with Flying W Airways from Bill Whitesell at Medford. Because of the massive air transport requirements for the construction of the Trans-Alaskan Pipeline to the North Slope oilfields, this company had large numbers of Curtiss C-46s and Lockheed Constellations. These airplanes operated in Alaska with Red Dodge Aviation titles as well as Red Dogs own airplanes including a DC-3, Grumman Goose, P-51 Mustang, and some Cessna Bobcats. During the 1960s,

Red Dodge operated five B-25 Mitchell fire bombers on contract with Alaska State authorities, enabling them to remain in service after the USFS banned B-25s in 1962 because of wing structural failures after retardant drops. The B-25s in service with Red Dodge were N3476G, N3521G, N7946C, N9857C and N9866C.

N3476G

This was a B-25J-15 with s/n 44-28932. After USAF storage at Davis Monthan, this B-25 was sold to Earl Dodge in June 1959. A civil registration of N3476G was issued in December of that year. In May 1962, a 1,000-gallon borate tank was installed, and she was operated as tanker #6. In September 1965, she was sold to Robert Schlaefli of Port Orchard, Washington.

N3521G

This was a B-25J-30 with s/n 44-86853. This airplane served only for a few months with the company. She was purchased in August 1959 from National Metals, Tucson, Arizona and was already sold to Blue Mountain Air Service, La Grande, Oregon in December 1959.

N7946C

This airplane was a B-25J-15 with s/n 44-28938. After this tanker had flown for some years with Wenairco, she was sold to Sports Air of Seattle, Washington in June 1966. A month later, she was again sold to Red Dodge Aviation and flew as tanker #4 in an overall light-yellow colour. In July of 1975, the B-25 was sold to Russell DeFrancesco and John Cahill of Cardiff, California.

Red Dodge Aviation's tanker #4 photographed at Merrill Field, Anchorage in September 1968. *(Neil Aird)*

N7946C in June 1971 at Anchorage without retardant tank, parked at the Red Dodge hangar office. She was painted overall light yellow with dark brown trim. The "Snoopy" badge on the nose is brown, black and white with black inscription "Happiness is a Thunderstorm". The lightning bolts are possibly red. (Geoff Goodall, b-25history.org)

Tanker #4 again in the left background at Merrill Field. The ship in the foreground is B-25J-20 s/n 44-29249. After the war, this airplane was transferred to Davis Monthan AFB and sold to Don Holm at Portland, Oregon, in June 1958. She was registered as N9637C. Later, she was withdrawn from use and was stored in the open field at Merrill Field. Finally, she was broken up at Merrill Field. Here she is photographed in her last days in September 1968. (Neil Aird)

N9857C

This was a B-25J-5 with s/n 43-28059. In September 1958, she was sold to Blue Mountain Air Service from La Grande, Oregon. There she was fitted with a 1,000-gallon tank in April 1959. In December 1959, she was sold to Earl Dodge. The company flew her as tanker #4 and sold her eleven years later in July 1970 to Edgar L. Thorsrud of Missoula, Montana.

N9866C

A B-25J-15 with s/n 44-28833. The B-25 was sold to Aero Spray Inc., Vancouver, Washington in October 1958. She was registered as N9866C. In May 1959, she was converted into a fire tanker with a 1,000-gallon tank. She was operated by Richardson's Airway as tanker #59. In September 1960, she was sold to Red Dodge where she flew as a tanker. The airplane crashed at Bremerton, Washington on 24 August, 1961. Her incomplete wreck was sold at an auction at Moses Lake in 1999.

N9857C photographed in September 1966. The company flew her as tanker #4 and owned her for eleven years. This airplane also had the nose art of the horse pulling the old-fashioned fire wagon. (Neil Aird)

N9866C with the same horse nose art at Merrill Field Airport in April 1961. The tank below the fuselage can be clearly seen. (Geoff Goodall)

L.K. ROSER, PHOENIX, ARIZONA

The L.K. Roser company of Phoenix, Arizona owned one B-25 air tanker. This B-25J was the last Mitchell of the block-30 series and was civil registered as N3337G. The bomber was purchased from the USAF disposal sale by Max Biegert in April 1959. In December 1959, she was sold to L.K. Roser. In June 1960, she was converted into an air tanker and was fitted with a 1,170-gallon tank. Roser flew the airplane for nearly five years as tanker #37C. In February 1965, the tanker was sold to Aero Industries of Addison, Texas.

Nice colour picture of N3337G of L.K. Roser somewhere in the early 1960s. The tanker is overall aluminium finished with white fuselage top and red engine nacelles. (Collection Dan Dinneen)

Sonora Flying Service, Columbia, California

In the 1950s, the U.S. Forest Service began using airplanes to fight fires. One of the early firefighting companies was Sonora Flying Service at Columbia, California. This company operated its firefighting airplanes under contract with the USFS in Arizona, California, and New Mexico. The firm also had a field and hangar at Silver City, New Mexico, with equipment for loading planes with fire blanketing bentonite slurry. Sonora Flying Service was the largest privately operated aerial forest firefighting firm in the West. Owner of Sonora Flying Service was Bob Roberts. Sonora had different airplanes like a Stearman biplane, TBM Avengers and a B-25. This was N9115Z. In 1958, this B-25J-20 was transferred to Davis Monthan AFB for storage. In January 1960, she was purchased by Sonora Flying Service for $2,000. In May 1960, she was converted to a fire-bombing tanker and the airplane could carry nearly 1,000 gallons of water or slurry. An upright tank in the centre of the bomb bay was divided into two compartments so that 465 gallons could be dropped at each pass over the fire front. She flew as a firefighting bomber aircraft #48. In October 1964, the B-25 tanker was sold to Sam Rawland and Morgan Hetrick at Osage, Missouri. The bomber ended her career in the U.K. as movie star in "Hanover Street" and in 1982, she went to the RAF Museum at Hendon, U.K. for display in the new Bomber Command Museum. She was repainted in an USAAF colour scheme.

N9115Z of Sonora Flying Service as tanker #48. The ex USAF code BD–366 is still visible on the metal finish. The picture was taken in the early 1960s. (Collection Dan Dinneen)

Sprayair Ltd., Scramento, California

A B-25 was operated for five years by Sprayair Ltd. at Sacramento. This bomber was a B-25J-25 with s/n 44-30721. She was stored at Davis Monthan AFB, Arizona in October 1957. In January 1958, she was purchased by Sprayair Ltd. and registered as N5455V. She was subsequently converted into an agricultural sprayer and fitted with spray bars under the tail of the plane. She was operated for five years by Sprayair. She flew as tanker #3. In March 1963, the airplane was sold to Carstedt Sales Corp. at Long Beach, California and one month later she went to Florida and was sold to Robert B. Becker in Miami. In August 1963, she was sold to Servicios Americanos S.A. in Miami. Finally, in December 1964, she was sold to Maravilla Inc. in Miami. In 1970 the aircraft disappeared; her fate is unknown.

N5455V was modified with a tail mounted spray bar and operated for five years by Sprayair Ltd. at Sacramento, California.
(Bill Larkins)

SPRUNG AVIATION, TUCSON, ARIZONA

Fred Russell Sprung was a hunter, horseman, aviator and enjoyed international travel. He enlisted in the Navy and trained as a pilot near the end of World War Two. Russell joined the Tucson Fire Department in 1949. As Asst. Chief he was involved in fighting some of the most horrific events in Tucson Fire history: the Supreme Cleaners gas explosion in 1963, the Pioneer Hotel fire in 1970 and the Copper State Chemical fire also in 1970. He founded Sprung Aviation in 1959, served as a Federal Aviation Administration flight instructor and flew converted B-25 bombers to fight forest fires in the western U.S. After retiring from Tucson Fire Department, he owned businesses including car washes and rental properties. In the 1950s, Sprung owned three B25J models; these were N3507G, N9451Z and N9452Z.

N3507G

This ship was a B-25J-30 with s/n 44-86843. She was stored at Davis Monthan in 1958. In 1959, she was purchased by Ace Smelting Co. from Phoenix, Arizona and registered as N3507G. In August 1959, she was sold to Sprung Aviation at Tucson. In July 1960, the bomber was converted into a fire tanker. Sprung sold the bomber to Major Air Corp. at Tucson in April 1962 where she flew as tanker #05C. In the late 1960s, the airplane was purchased by Tallmantz and used in the film "Catch-22".

Tanker #05C of Major Air Corporation in the mid-1960s. The airplane, registered as N3507G, had been equipped by Sprung Aviation with a 1,000-gallon tank in July 1960. In April 1962, Sprung sold the airplane to Major Air Corp. The colour picture was taken at Ryan Field in April 1967 while still carrying the tank. (Collection Wim Nijenhuis, Renee Francillion)

N9451Z

This airplane was also stored at Davis Monthan in 1958. She was a B-25J-25 with s/n 44-30493. In January 1960, she was sold to National Metals Inc. at Phoenix, Arizona and registered as N9451Z. In August 1960, she was sold to Sprung Aviation and planned for conversion into a tanker. But the airplane was stored until July 1968, when she was purchased by Tallmantz Aviation at Santa Ana, California. She flew in the movie "Catch-22" and was later modified as a camera ship.

N9452Z

Like N9451Z, this B-25J-25 with s/n 44-30649, was stored at Davis Monthan in 1958 and sold to National Metals Inc. in January 1960. She was registered as N9452Z. In August of that year, she was sold to Executive Management & Investment Corp. at Tucson. Four years later in August 1964, she was purchased by Sprung and one year later in August 1965, she was sold again to Don Underwood of Donaire Inc. at Phoenix, Arizona. In June 1968, she was sold to Tallmantz Aviation, Orange County for use in the film "Catch-22".

N9451Z on the ramp at Orange County in 1968. This airplane pretty much looks the same as it did in 1960, when she was purchased by Sprung for conversion into an air tanker, something that never happened. Tallmantz bought the airplane in July 1968 on behalf of Filmways, Inc. getting ready to film "Catch-22". (www.warbirdregistry.org)

EDGAR L. THORSRUD, MISSOULA, MONTANA

In the mid-1950s, Edgar L. Thorsrud formed his own company at Missoula, Montana. Ed Thorsrud was born on 5 August, 1922, in Watford City, North Dakota, son of Norwegian immigrants. In 1927, the Thorsrud family moved to Missoula. His Norwegian skiing heritage led him to explore the Montana alpine long before the advent of ski lifts. During World War II, Thorsrud flew for the Army Air Corps ferrying troops and cargo from England to the Continent. After the war, he returned to Missoula and flew for the Johnson Flying Service delivering cargo, smokejumpers, and fire fighters to fires throughout the West. Flying was his passion and

his livelihood. He was a pioneer in the early days of mountain aviation, flying various planes, including Tri-Motors and DC-3s, to remote mountain airstrips. He delivered essential cargo to the backcountry and transported smokejumpers and fire fighters to fires throughout the West. In the mid-1950s, this experience led him to invest in his own slurry planes, including a TBM and B-25. He dropped retardant for years in Alaska, Montana, and Idaho, taking him away from his family for long absences during the fire season. Edgar L. Thorsrud flew with two B-25s, registered as N9857C and N9088Z.

N9857C

This was a B-25J-5 with s/n 43-28059. In 1959, she was fitted with a 1,000-gallon tank. Red Dodge Aviation Inc., Anchorage, Alaska flew her as tanker #4 and sold her in July 1970 to Edgar L. Thorsrud. He flew the B-25 for nine years and sold the airplane in December 1979 to Max Power Inc. at Carlsbad, California.

N9088Z

The B-25J-25 with s/n 44-30733 was originally delivered in 1945. In December 1959, she was sold surplus to National Metals from Phoenix, Arizona and registered as N9088Z and immediately sold to Johnson Flying Service of Missoula, Montana. In December 1966, she was sold to Ed Thorsrud and converted into a fire bomber. She flew as tanker #8Z, based in Alaska. Thorsrud would travel to Fairbanks every fire season to fly the Mitchell for the Alaska Department of Natural Resources. On 27 June, 1969, the B-25 was used to fight the large Manly Hot Springs fire. Then she suffered a double engine failure shortly after take-off. The pilot was forced into a wheels up landing on a sandbar in the Tanana River. Two days after the crash, Thorsrud and his mechanic removed the engines, wheels, and control surfaces that could be used on another B-25 that he was going to buy. The cost of recovering the airframe was more than the price of another B-25 at the time. The reaming airframe of the "Sandbar Mitchell" was left

N9857C stored at Fairbanks in September 1968. The engines have been removed. At the time, she was owned by Red Dodge. Two years later in July 1970, she was sold to Edgar L. Thorsrud. (Neil Aird)

Tanker N9088Z in action.
(Warbirds of Glory Museum)

derelict at the crash site and over the years with parts being removed by scavengers and other restorers. In July 2013, she was recovered by the Warbirds of Glory Museum and transported to the museum in Michigan. Currently, the bomber is undergoing restoration by the museum.

TRANS-WEST AIR SERVICE / AERIAL APPLICATORS INC., SALT LAKE CITY, UTAH

Below: *Edgar L. Thorsrud flew with N9088Z which was numbered tanker #8Z. (Warbirds of Glory Museum)*

Aerial Applicators Inc. was a fire-bombing operator based at Salt Lake City Airport, Utah. It was a subsidiary of Trans-West Air Service with John D. Streeter as Managing Director of both companies. It had a fleet of Grumman TBM Avenger tankers, Douglas DC-6s and DC-7s and also a B-25. The company ceased its operations during 1981 and the remaining aircraft were sold to other tanker companies. The B-25 was a J-15 model with s/n 44-28925. This airplane was originally assigned to the 380th Bomb Squadron, 310th Bomb Group in Italy in 1944. She was declared surplus in January 1958. She was sold to Parsons Airpark at Carpenteria, California and registered as N7687C in July 1958 after conversion into an agricultural sprayer. In 1960, extensive modifications were made with the installation of electron-

The former combat ship now in a friendly civil role on the ramp at Salt Lake City in July 1966. At this point, the airplane was owned by Trans–West Air Service. (Milo Peltzer)

Below: *The same ship in her yellow colours photographed a month later at Salt Lake City. (Neil Aird)*

ic test equipment. By August 1964, she was sold to Aerial Applicators Inc. and operated by Trans-West Air Service. Nearly four years later in May 1968, she was purchased by Tallmantz Aviation of Santa Ana, California for use in the filming of "Catch-22". Currently, the airplane is owned by the Cavanaugh Flight Museum, Addison, Dallas, Texas and is named "How 'Boot That!?" just like she had in the war.

E.D. Weiner, Los Angeles, California

Very little is known about the B-25s of E.D. Weiner. In the late 1960s, Ed Weiner had two racing P-51 Mustangs and competed in different air races. In 1969, Weiner had a heart attack during an air race and died a week later. The B-25s of Weiner were N9865C and N9899C.

N9865C
This was s/n 44-28834, a B-25J-15. This airplane was converted into a fire tanker in May 1960. After she had several different owners, she was purchased by Ed Weiner in March 1963. She kept the same tanker number 30 that she had when she was owned by Western Air Industries at Redding. In July 1963, the airplane was sold to Aero Enterprises, La Porte, Indiana, and the retardant tank was removed.

N9899C
This was a B-25J-20 with s/n 44-29127. This airplane was stored at Davis Monthan in December 1957. She was bought by E.D. Weiner in September 1958 and registered as N9899C. After four years, she was sold to Constatine Zaharoff, Grand Prairie, Texas. in September 1962.

E.D. Weiner owned tanker #30 for a few months in 1963. She kept the same colours and tanker number from her former owner Western Air Industries at Redding, later renamed Aero Union Corporation.

(Collection Wim Nijenhuis)

WENAIRCO INC., WENATCHEE, WASHINGTON

Wenatchee Air Service of Wenatchee, Washington was an agricultural spraying service company. The name Wenatchee Air Service Inc. changed to Wenairco Inc. on 20 September, 1960. The company also operated Wenairco of Canada Ltd., Vancouver, British Columbia. The fire tanker operations of Wenatchee Air Service, respectively Wenairco Inc., were taken over by William A. Dempsay, Rantoul, Kansas. Dempsay also flew fire tankers and he continued Wenairco in fire bombing under the name Central Air Service Inc., East Wenatchee, Washington, later Maricopa, Arizona. Dempsay was widely known from his long experience in the insecticide and chemical spraying business with his family's Kansas based companies established by 1961.

Wenairco was located at Fancher Field. This was an airfield located in Douglas County, Washington, named after Major John T. "Jack" Fancher, a World War I veteran. The City of Wenatchee operated Fancher Field until 1949, when it was turned over to Wenatchee Air Service and used for pilot training, spraying operations, and other general aviation. Fancher Field was evidently closed at some point between 1964-1966. However, the airfield continued to operate in some capacity after it was no longer depicted on aeronautical charts. It gradually fell into disuse over the next few decades. The site was the home of Wenairco until it's closing in 1985. Today, Fancher Field's historic runways have been turned into a housing development.

The company Wenairco had two PV-2 Harpoons, a PB4Y-2 Privateer and three B-25s. They were used as fire bombers. The three B-25s were and N5277V, N7946C and N9613C.

N5277V/CF-OND

William E. Boeing/Aero Boeing, Seattle, Washington, bought this airplane in January 1958 and the civil registration N5277V was assigned. In July 1960, she was sold to Wenatchee Air Service in Wenatchee, Washington. From January 1964, she was operated by Wenairco of Canada Ltd., Vancouver, British Columbia. There she was civil registered as CF-OND and she flew as tanker #38.

This is CF-OND at the time the ship was owned by Wenairco and flew as tanker #38. Wenairco sold her to North Western Air Lease in 1970. (Collection Wim Nijenhuis)

Crew members are filling the water bomber in 1960 at Cranbrook, British Columbia, Canada. This is N7946C. Skyway Air Services Ltd. of Langley, British Columbia operated for the British Columbia Forest Service. Skyways hired the Wenatchee B-25 from Washington State to cover the Cranbrook area. (City of Vancouver Archives)

Probably, because the same month N9613C of Wenairco, that also flew as tanker #38, crashed near Twin Wasp, Washington. In 1970, CF-OND was sold to North Western Air Lease in St. Albert, Alberta and operated as tanker #90.

N7946C

This was a B-25J-15 with s/n 44-28938. On 18 January, 1958 she was sold from the USAF to P.J. Murray of Oxnard, California and a civil registration of N7946C was issued. She was again sold in October 1958 to Wenatchee Air Service in Wenatchee,

Washington. In May 1959, a 1,200-gallon tank was installed, and the airplane was used as a fire bomber. She flew as tanker #39. She was removed from service in May 1965 and her registration was cancelled. In June 1966, her registration was reinstated when she was sold to Sports Air of Seattle, Washington.

N9613C

The other ship was s/n 44-30377 a B-25J-25. This airplane was withdrawn from use and stored at Davis Monthan in 1957. In 1958, she was sold to Arrow Sales Inc., North

The nose with the red letters Wenairco of N7946C at Cranbrook. (City of Vancouver Archives)

Hollywood, California and registered as N9613C. In June 1958, she was purchased by Wenatchee Air Service. In August 1958, she was converted into a sprayer/fire tanker and flew as tanker #38. On 26 July, 1960, the airplane crashed after suffering structural failure while fire bombing, near Twin Wasp, Washington.

On the nose of N9613C the remnants of the old USAF code BD-377 are still visible. (City of Vancouver Archives)

Left: *Water bomber N9613C is waiting to be filled at the Cranbrook Airport in 1960.*
(City of Vancouver Archives)

MAINTENANCE, RESTORATION AND TRADING

In most cases, the post-war military airplanes intended for civil use had to be converted for this purpose. In addition, adjustments were necessary due to legislation and regulations. Of course, these outdated airplanes demanded quite a bit of maintenance. This meant that companies arose that specialised in the conversion and maintenance of post-war military airplanes and in their trade. That also applied to the B-25. And there were other companies that further specialised and focused on full or partial restoration of B-25s. Some of the most important are described below.

AERO TRADER, CHINO, CALIFORNIA

Together with Tom Reilly (Kissimmee, Florida/Douglas, Georgia), Aero Trader is considered the world's most famous B-25 restorer. They have had several B-25 projects under control and delivered beautifully restored B-25 warbirds. Aero Trader is specialised in the restoration of military aircraft from World War II and the period immediately after, primarily piston engine types. The company will also train people to fly them or rent a B-25 camera ship and crew for air-to-air photography sessions. But the name Aero Trader has become especially synonymous with B-25 restorations. Carl Scholl and Tony Ritzman operate Aero Trader from their facility in Chino, California. The business began in 1976, when Carl Scholl bought a derelict B-25 parked at Ramona Airport, California. He had always been interested in aviation and buying the B-25 seemed like a good idea to him. This airplane was registered as N3155G.

At this same time, Scholl met Joe Davis who had also purchased a derelict B-25 at a small airfield near Omaha, Nebraska. Both Davis and Scholl began looking for B-25 parts, and Scholl discovered the surplus airplane parts business owned by Jack Hardwick in El Monte, California. Scholl and Davis went in together to purchase Hardwick's entire B-25 inventory and formed a corporation called the Historic Aircraft Preservation Group. Tony Ritzman joined them in July 1979 and then they were slowly drawn into the aviation world of warbirds. For the exchange and sale of warbird parts, they used the new company name Aero Trader.

They moved to Borrego Springs, California and this became the new home of Aero Trader. The company began to attract other B-25s from around the area for mechanical work or storage. By 1982, Scholl and Ritzman were both type-rated in the B-25 so they could fly their own B-25s. Over the next several years, Aero Trader purchased a number of derelict or near-derelict airframes and several substantial parts inventories. By 1984, Aero Trader had a lot of B-25s and parts from which they could run their business. Scholl and Ritzman were hired by air

Sign at the hangar of Aero Trader at their home base at Chino. (Steve Nation)

museums and others around the country who needed to have a complete B-25, or just an engine, reconditioned. They also had on microfilm the original blueprints from the builder of the B-25, North American Aviation. In 1985, the company moved its day-to-day operations back to Chino to become more customer accessible. A hangar,

offices, and a small storage area were obtained. Much of the collection of parts and airframes remained at Borrego Springs, but ongoing restoration projects were moved to the new facility at Chino. Aero Trader's specialty remained the reconstruction of tired B-25s and other warbirds into pristine examples of the types.

Through the years since the early 1980s, Aero Trader has anchored much of the B-25 work in the warbird world. Aside from their restoration efforts, Carl Scholl and Tony Ritzman have had their hands in several notable projects involving B-25s. In 1987, Ritzman joined John Crocker in flying Steven Grey's B-25D, N88972, across the Atlantic for delivery to The Fighter Collection at Duxford, England. See chapter United Kingdom/The Fighter Collection. In 1989, Tony was one of the pilots who flew the B-25 camera ship N1042B for the filming of "Memphis Belle" in England. But, the primary market for Aero Trader's work is focused on the B-25 owners who bring their aircraft in for restorative or mechanical work specifically suited to the state of the aircraft and the budget of the owner and the company also has expertise in other piston and jet-powered warbirds.

Today, Aero Trader continues to provide quality restoration and maintenance services to a large warbird ownership population. The company continues to maintain the largest inventory of B-25 airframes and parts in the world, which also serve as the basis for future projects.

N201L at Chino in July 1987 as "Bat Out of Hell" and in March 1990 after her glass nose was fitted again.

(Collection Wim Nijenhuis, Tom Tessier)

Some Aero Trader B-25 projects

In the course of time, Aero Trader has done maintenance and restoration work on civil B-25 Mitchells like: N25NA, N201L, N325N, N345BG, N898BW, N1943J, N2825B, N3155G, N3438G, N3442G, N3476G, N5548N, N5672V, N7674, N7687C, N8195H, N9856C, N10564 and N88972.

Below are described a few outstanding B-25 projects that have been taken over by Aero Trader in the past.

N201L

The B-25J-25 with s/n 44-30606 now known as "Tootsie", was registered as N201L. By December 1970, she was sold to John Bishop of Hobbs, New Mexico and was restored with the assistance of Aero Trader. She was fitted with a gun nose and flew at that time as "Bat Out of Hell". The maintenance and restoration work were done by Aero Trader. She was sold again in March 1989 to Ted and Sharon Kay Melsheimer of Carson City, Nevada. She was again restored by Aero Trader and received her current markings and flies in blue colours as "Tootsie". In December 2012, she was sold to TSM Enterprises LLC at Carson City, Nevada.

N325N

After a three-year restoration, the company Aerocrafters flew with the solid nose B-25 "Sunday Punch", carrying the civil registration N325N. The airplane was sold to Fagen Fighters WWII Museum in May 2012. She was restored by Aero Trader in 2012-2013 and fitted with a glass nose. She is now flown in a beautiful Olive Drab and Neutral Grey scheme and named "Paper Doll".

N345BG

This is a B-25J-30, s/n 44-86777. In April 1997, this ship with the former registration N9167Z was sold to Bud Firth and moved to Baton Rouge, Louisiana. In 1999, the ship was purchased by Dave Wheaton of Sand Springs, Oklahoma. She was re-licensed and returned to airworthy status by Aero Trader and she flew as "Martha Jean". In 2011, she was purchased by the Liberty Aviation Museum in Port Clinton, Ohio. The B-25 underwent comprehensive restoration work at Aero Trader over the winter of 2011/2012. She is now flying as "Georgie's Gal".

N201L now as "Tootsie" in June 2008, at the Annual Carson City Airport Open House.
(Gary Schenauer)

N325N was flown by Aerocrafters as "Sunday Punch". In 2012–2013, she was restored by Aero Trader to become "Paper Doll" with a greenhouse nose.
(Fagen Fighters WWII Museum)

The overall natural aluminium finished "Martha Jean" at the Biplane Expo in Bartlesville, Oklahoma in 2009. In 2012, the ship received an Olive Drab paint and named "Georgie's Gal". (anyjazz65)

This is the result. "Paper Doll" as she flew during the Second World War with the 321st BG, 447th BS at Solenzara, Corsica. The aircraft was delivered to the unit in bare metal finish. The Olive Drab camouflage on the upper surfaces was applied in late May 1944. The propeller hubs were painted blue. (Fagen Fighters WWII Museum)

"Apache Princess" during engine tests at Fantasy of Flight in Polk City, Florida in 2012. (Fantasy of Flight Museum)

N1943J

In May 1981, Aero Trader purchased a B-25J registered as N26795. She underwent an extensive authentic restoration and was sold to Kermit Weeks in Tamiami, Florida in May 1983. Her current registration of N1943J was assigned in July 1984. She was rebuilt to flying condition and made her first flight in March 2000, following a "100 per cent" accurate restoration by Aero Trader. She was flown as "Apache Princess", a tribute to the original airplane of the 501st BS, 345th BG "Air Apaches" in the South West Pacific Area. Nowadays, she resides at the Fantasy of Flight Museum in Polk City, Florida.

N2825B

This is the oldest known B-25 in existence. Aero Trader has done substantial work on this B-25 for several owners, including the addition of a glass nose, reworked fuel system, and cockpit work. The airplane owned by Jeff Clyman, currently flies as "Miss Hap". She was transferred in December 1983 to Fighting Air Command from Dallas, Texas where her modified solid nose was replaced with a standard greenhouse nose. At the time, she was operated as "Proud Mary". She was sold in October of 1989 to Courtesy Aircraft from Rockford Illinois and then to Jeff Clyman/TBF Inc. of Tenfly, New Jersey. As of

Right: *The splendid nose art of N1943J. (Collection Wim Nijenhuis)*

"Miss Hap", the oldest known B-25 in existence during the B-25 Gathering at Grimes Field in April 2017. The Grimes Gathering of B-25s led to a tribute to the Doolittle Tokyo Raiders and the 75th Anniversary of their raid on Japan, held at the National Museum of the U.S. Air Force, near Dayton, Ohio. Aero Trader has done substantial work on this B-25 for several owners (Wim Nijenhuis)

This is N3155G in October 1978 at Chino, for a long-term restoration with Carl Scholl.
(Phillip Dawe)

2002, she has been operated by the American Airpower Museum in Farmingdale, New York and flown as "Miss Hap".

N3155G

This B-25 was Carl Scholl's original airplane purchased in 1976. The B-25J-25 was in Chino for a long-term restoration. In June 1980, she was sold to Don Davis of Casper, Wyoming. Nowadays, she is owned by Claire Aviation Inc. and is currently flown out of Pennsylvania as "Take-Off Time".

N3438G

This airplane was recovered in a semi-derelict condition from Turlock, California. In August 1983, Aero Trader bought N3438G and ferried her to Borrego Springs for a rebuild. In February 1985, after restoration, she was sold to Wiley Sanders Truck Lines in Troy, Alabama. She first flew as "Samantha", then later as "Ol Grey Mare". In June 2004, she was purchased by Hans Lauridsen of Glendale, Arizona. She currently resides at the Lauridsen Aviation Museum at Buckeye Airport in Buckeye, Arizona.

N3438G "Ol Gray Mare" of Wiley Sanders at Oshkosh in 1993. (Collection Wim Nijenhuis)

N5672V

The B-25 N5672V switched a couple of times in Canada and eventually ended up back in the U.S.A. with Aero Trader in 1993. She was removed from service and sold in June 1994 to C&P Aviation Services, Blaine, Minnesota. Her registration was changed to the current N5672V in September 1994. In 1994, a restoration was started to create "Betty's Dream". From 1994 to 1999, she underwent restoration work. After restoration, she flew for the first time in March 1999, including her bomb bay doors which had been removed when she was a fire bomber. Since restoration, she has flown as "Betty's Dream". In 2011, she was operated by Fighter Hangar 1, now known as the Texas Flying Legends Museum, Houston, Texas.

N7674

In early 1978, N7674 was purchased by Scholl and Ritzman from long-time warbird parts supplier Bob Sturges. N7674 was an ex-RCAF Mitchell then parked at the Troutdale, Oregon airport. Scholl drove a replacement engine to Troutdale and got the bomber ferried back to Chino. She was restored to airworthy condition and a Certificate of Airworthiness was issued in August 1978. In March 1979, the airplane was sold to Wings & Wheels, Burlington, North Carolina, providing needed capital for the company. About one year later, she crashed in Colombia.

Restored by Aero Trader in the 1990s and painted with memorable colours and great nose art, N5672V "Betty's Dream". (Collection Wim Nijenhuis)

N7687C

After this B-25 had been displayed on a pole at Forest Lawn Gardens Veterans Cemetery, Pittsburgh, Pennsylvania, she was sold to Doan Helicopter of Daytona Beach, Florida in August 1984. Her registration N7687C was reinstated. In October 1984, she was trucked to Florida for a rebuild, but stored disassembled at Kissimmee, Florida. In November 1992, she was sold to the Cava-naugh Flight Museum, Addison, Dallas, Texas. The museum restored the B-25 with the help of Aero Trader in 1995. The restoration of the bomber was complete in every detail and all the plane's systems are fully operational. She is the pride and joy of Aero Trader's efforts.

Bob Sturges' N7674 at Troutdale airport. She went to Aero Trader and crashed later in Colombia. (Jack Cook)

N7687C "How Boot That?!" at the Cavanaugh Flight Museum, Addison, Dallas, Texas in 2008. (Perry Quan)

N8195H

Aero Trader's initial mechanical work was completed for N8195H of Mike Pupich at Borrego Springs to get this B-25J re-licensed. The bomber flew as "Heavenly Body". Mike S. Pupich of San Fernando, California bought the old bomber in 1972 and has lavished work and money on her to restoring her, flying her in air shows and trying to turn her into an exact copy of the "Heavenly Body," a B-25 flown in the China-Burma-India Theatre during the war. She was restored to airworthy status in 1974. In April 1992, she was the first of two B-25s that launched off the aircraft carrier USS Ranger to commemorate the 50th Anniversary of the Doolittle Raid. In January 2014, she was sold to the Erickson Aircraft Collection, Madras, Oregon.

September 1977, N8195H is parked at Van Nuys, California and was at the time a clean ship with markings of the post-war USAF. (Geoff Goodall)

"Heavenly Body" at Nellis AFB, Las Vegas, Nevada in November 2011. She is generally considered as one of the best restored Mitchells. (Tomás Del Coro)

"Pacific Princess" in her U.S. Marines colours in the 1980s and without the blue paint at the Confederate Air Force Show at Midland, Odessa in October 1994. (NARA, Wim Nijenhuis)

N9856C

In May of 1973, N9856C of Tallmantz Aviation at Santa Ana was sold to Ted Itano. She was used as a static prop in the filming of "1941" at Long Beach Airport. She was restored by Aero Trader. In April 1992, she led the formation of B-25s in a "Missing Man" formation to honour the 50th Anniversary of the Doolittle raid. In both August and October of 1995, she launched from the aircraft carrier USS *Carl Vinson* to commemorate the 50th Anniversary of the end of World War II. In 2000, she took off from the deck of both the USS *Lexington* and the USS *Constellation* for the movie "Pearl Harbor". Aero Trader worked out an agreement to maintain the B-25 in return for flight privileges. In 2005, the airplane was purchased by Carl Scholl, Tony Ritzman and Bruce Graham. Currently, she still flies as "Pacific Princess" by B-25 Mitchell LLC, Missoula, Montana. The airplane was initially painted in the three-tone U.S. Marines paint scheme with nose art. In the early 1990s, the paint was removed, and she was overall natural aluminium finished but the nose art was retained.

N10564

This was an ex-"Catch-22" aircraft and in the early 1980s operated by John F. Marshall of Ocala, Florida. He bought the airplane from Wings of Yesterday Museum, Santa Fe, New Mexico and flew her to Williston Airport, near Ocala, Florida, in November 1979. Extensive restoration work and interior equipment installation was performed by Aero Trader. Thereafter, Marshall flew the airplane as "Little Brown Jugs" and later "Carol

"Carol Jean" at the CAF Air Show at Harlingen, Texas in 1983. The nose art is based on the illustration "Torches at Midnight" of Aberto Vargas.

(Alex Christie, Collection Wim Nijenhuis)

Jean" in honour of his wife. Marshall continued to fly the bomber and it appeared regularly at air shows. In 1985, Marshall read of the National Air and Space Museum's interest in acquiring a B-25 and decided that the bomber was ready for permanent re-

tirement. On 18 November, 1985, Marshall landed at Dulles Airport and taxied "Carol Jean" into Smithsonian custody.

N88972

Aero Trader did mechanical and restorative work on the old B-25D N88972. This work included new engines, propellers, interior finishing, glass, and a turret and was done for the Fighter Collection in Duxford, England. In 1987, the airplane was moved to Aero Trader in Chino for restoration. After restoration, she flew to England and arrived at Duxford in November 1987. She was painted with the RAF serial KL161 and aircraft code VO-B of No. 98 Squadron and named "Grumpy". The airplane was displayed and flown for several years with The Fighter Collection at European air shows.

GRAND CENTRAL AIRCRAFT COMPANY, GLENDALE, CALIFORNIA

A beautiful B-25 was used by Grand Central Aircraft Company at Glendale, California. This was a B-25H-1 with s/n 43-4336. Grand Central Aircraft Company was a corporation which was engaged in the production and repair of aircraft equipment at Glendale, California, and Tucson, Arizona. In 1950, Grand Central Aircraft Co. leased two hangars at Tucson Municipal Airport in Arizona to rebuild the larger aircraft required for the Korean War. Jets and large aircraft were sent to the Grand Central Service Centre in Tucson. Medium bombers and fighters were sent to Glendale. Grand Central (at both locations) was recognized as the largest repair, overhaul, and modification station in the country. Thousands of P-51

Mustangs, C-47 Dakota's and other aircraft transitioned through Grand Central Airport for updating and refurbishment. Also, many of the post-war B-25s were updated and refurbished by Grand Central Aircraft at Glendale. Accounts of all these airplanes included the Nationalist Chinese Air Force, the Brazilian Aeronautical Commission, Uruguayan Air Mission, Standard Oil, Sinclair Oil and numerous airlines.

In 1950, the Curtiss-Wright Technical Institute became part of the Grand Central Aircraft Company and the institute was renamed as the Cal-Aero Technical Institute. This institute was an early professional trade school operated by the Curtiss-Wright

Corporation for airplane maintenance training. The U.S. Air Force used the institute to train mechanics on contract until 1952. Enrolment dropped sharply after the cancellation of the contract and the facility closed in 1954. In 1954, Grand Central Aircraft was the city's largest employer at the time. It also played a role in the related field of rocketry, when in 1955 the Grand Central Rocket Company was established. The company tested solid rocket propellants among the old revetments around the field. Although Grand Central Aircraft was still the city's largest employer, rising taxes, declining business, and pressure to close the airport and convert the property into a large industrial area resulted in the closure of the Grand Central Aircraft Company in 1959.

The B-25 used by Grand Central Aircraft Co. was purchased from Cal Aero Technical Institute, Glendale in August 1950. She was a B-25H-1, s/n 43-4336, registered as N67998.

1949, B-25s on the platform of the Grand Central Airport. In the picture are apart from P-51 Mustangs and a DC-3, five B-25Js. The B-25s are overall natural aluminium finished and without the rudders and markings. (Don Ayres)

By July 1951, she was converted for civil use and fitted with a glass nose. She was modified with a B-25C tail stinger and a passenger interior. In June 1957, the glass nose was faired over, and the waist windows were modified. In July 1957, she was registered as N96GC. In October 1961, the plane was sold to Wheeler & Ryan, Fort Smith, Arizona and one year later she became property of Bert Wheeler at Fort Worth, Texas. In June 1964, she was sold to Air Services Inc., Addison, Texas. There she was modified with radar, gun cameras and a nose radome. Finally, after she had a few other owners, she crashed during a drug run into the mountain and was destroyed at Dawsonville, Georgia on 3 June, 1975.

Centre: *The beautifully modified N96GC of Grand Central Aircraft Co. at Newark, New Jersey in 1960.*

Bottom: *The same ship but now in service with Wheeler & Ryan Oil and Gas Exploration. The company name with a globe is painted on her nose.* (Earl Holmquist, Collection Wim Nijenhuis)

HAYES AIRCRAFT CORPORATION, BIRMINGHAM, ALABAMA

The company that would become known as Pemco started in Birmingham, Alabama, in 1951 as Hayes Aircraft Corporation. The company refurbished military airplanes and, during its heyday, employed some 12,000 workers. Hayes started operations in a defence production facility adjacent to the Birmingham Municipal Airport. It began its journey into the aircraft maintenance business with the Air Force Material Command under contracts to perform aircraft maintenance, modification, and repair projects. Between 1952 and 1954, about 1,000 B-25Js went through the Hayes Aircraft Corporation for IRAN (Inspect and Repair As Needed). These airplanes were equipped with such features as an automatic pilot, bomb bay fuel tanks, AN/ARN-14 radio gear, dual UHF/VHF, and demand oxygen systems.

inders of each engine. In 1953/1954, Hayes modified additional B-25Js as TB-25Ns. They were like the preceding TB-25Ls but were fitted with R-2600-29A engines.

Hayes was located in a 1.8 million-square-foot complex, minutes from downtown Birmingham. The hangar has 10 bays, each 160 feet wide and 725 feet deep, with 40-foot ceilings and concrete floors a foot thick. The complex has a paint hangar, warehouse space, offices and more. The commercial facility operated by Hayes and later by Pemco Aeroplex, has a long history in aviation, serving as a centre for maintenance, repair and overhaul operations for decades. When Pemco operated the facility a decade ago, its workforce numbered 2,300 and it had major contracts with the military for work

on C-130 Hercules transport and KC-135 tanker aircraft. Pemco Aviation Group Inc. performs a variety of aerospace-related work through its three main subsidiaries. Pemco Aeroplex, Inc., in Birmingham, Alabama, has been maintaining and modifying KC-135 tankers and Hercules C-130 transports since the 1960s. Pemco World Air Services, Inc., in Dothan, Alabama, performs heavy maintenance for commercial clients. Pemco Engineers, in Corona, California, specialises in cargo handling systems. Military work accounts for about two-thirds of the company's business. In September 2007, Pemco changed its name to Alabama Aircraft and in 2011, the company was bought by Kaiser Aircraft. Currently, the facility is leased by Stewart Industries International.

August 1956, employee Mr. Walter Templin is standing in front of the Hayes building with B-25s in the background.
(Ed Clark/Life)

Under contract with Hayes, B-25Js were modified for specialised advanced pilot training under the designation TB-25L. The airplanes were rebuilt from the ground up. All armament and armour were removed, and the pilot's three-piece windshield was replaced with a one-piece windshield that was equipped with wiper blades and an anti-icing system. The front engine hatch was enlarged. Two passenger seats were added forward of the bomb bay, and five seats were installed in the aft fuselage. On some airplanes, exhaust semi-collector rings replaced the "S" stacks on the top seven cyl-

After the war, many B-25s were modified by the companies Hayes and Hughes. In 1951, the Hayes Aircraft Corporation started in Birmingham, Alabama. The company refurbished military aircraft and, during its heyday, employed some 12,000 workers. The first major contract acquired by Hayes was to recondition and convert 125 B-25Js. This is the front page of the Hayes News of 15 July, 1955. The caption under the picture says: "They pioneered Hayes Aircraft 'plane delivery. Of the original Hayes Aircraft mechanics and technicians who prepared 125 stored B-25's for flight from Pyote, Texas, in 1951, the above 25 men represent part of those still at Hayes." (Facebook.com)

HUGHES AIRCRAFT COMPANY, CULVER CITY, CALIFORNIA

The company of Howard Hughes owned three B-25 Mitchells for executive transport. In the 1950s, Hughes was another company that modified a large number of B-25s for post-war training for the USAF. Hughes Aircraft Company was established under the umbrella of the Hughes Tool Company. Howard Robard Hughes, Sr., an early Texas oilman, solved the problem of drilling for oil through rock by developing a drill bit with cone-shaped revolving cutters and steel teeth capable of pulverizing rock. He patented his inventions in 1909, and the Hughes Rock Bit revolutionized the well-drilling process. Initially, the Houston-based company had almost a monopoly in the business and amassed enormous revenues and profits. Even after key patents expired in the 1930s and 1950s, it continued to dominate the industry until the 1970s. In 1909, Hughes and Walter B. Sharp formed the Sharp-Hughes Tool Company and opened a plant in Houston to manufacture the bit. When Sharp died in 1912, Hughes took over management of the company, which he renamed Hughes Tool Company on 3 February, 1915.

In 1918, Hughes purchased the remaining company stock, which had been sold by Sharp's widow to Ed Prather. Hughes continued to improve on his bit and maintained a near monopoly in bit technology by patenting every part of the bit and by buying the patents of competitors. He held some seventy-three patents when he died in 1924 and his son, Howard Robard Hughes, Jr., took over control of the company. Under his direction, the company became a multibillion-dollar enterprise and ventured into the motion picture business via Hughes Productions during the 1920s, into the airline business in 1939 with the acquisition of a controlling interest in Transcontinental and Western Air (later renamed Trans World Airlines) and into real estate, hotels and gambling in Nevada. In the 1930s, he pursued his interest in aviation by forming the Hughes Aircraft Company. The Hughes Aircraft Company was originally formed in 1932 under the control of the Hughes Tool Company.

With the money he made as the head of Hughes Tool Company, Howard Hughes moved to Hollywood to begin his career as a film director, an entirely self-funded venture. Hughes Aircraft, as a division of Hughes Tool Company, kept track of the expenses of his personal interest in flying and developing airplanes. Hughes established himself as a celebrity test and racing pilot, setting multiple speed records. The Hughes Aircraft Company won defence contracts during WWII to develop high speed pursuit and reconnaissance airplanes and communication systems. In 1941, the company moved from Glendale to Culver City, where it established two facilities, including a plant and an airport on a tract of land along Ballona Creek.

At that time, there were about 500 employees, 100 of whom were engineers, and Hughes needed a facility to help him produce his designs for the military. During World War II, a Hughes engineer developed a flexible feed chute to speed the operations of machine guns on B-17 bombers. The company also made electric booster drives, making machine guns less likely to jam, and manufactured wing panels and other parts for military planes from Du-

ramold – the patented wood substitute for metal. Most of the prototypes the company produced never saw production. At the Culver City plant, Hughes developed two of his well-known projects: the HK-1 (Hercules) Flying Boat, better known as the 'Spruce Goose', and the XF-11 reconnaissance plane. Although Hughes Aircraft originally produced airplanes, after WWII the company transferred its attention to helicopter production. For almost half a century, Hughes Aircraft Company was a major player in the design, development, and production of high technology systems for scientific, military and commercial applications. In the 1950s, Hughes of Culver City modified approximately 150 B-25Js for bomb and fire training purposes, they were designated as TB-25K and TB-25M. The TB-25K was the designation given to B-25Js converted as trainers for the operators of the E-1 radar fire control system and TB-25M for B-25Js with installation of the more advanced E-5 fire control system.

In the early-1970s, Hughes Tool took over the largest regional air carrier in the western United States: Air West, renamed Hughes Airwest following the purchase. Hughes

Tool also briefly owned Los Angeles Airways, a small airline operating a commuter service with a fleet of helicopters. The diversification of the company caused the expansion of corporate offices, research labs, and manufacturing plants. From the 1950s through the 1980s, Hughes Aircraft acquired facilities in El Segundo, Hawthorne, Fullerton, and San Diego, California, and in Tucson, Arizona. By 1983, the company was the largest industrial employer in California, and the largest employer in the Los Angeles area. It had become the nation's largest defence electronics contractor, and the seventh largest overall Pentagon contractor. Hughes remained the sole owner of Hughes Tool division, which had long since become a part of a holding company of the same name, until 1972, when he sold the division's assets to a group of his major employees. The company, keeping its original name, then went public in a stock sale for $150 million. Hughes Tool Company merged with Baker International to

Around the fifties and sixties of the last century, the Hughes Tool Company owned three B-25s, N3513G, N3968C and N75831.

N3513G

In August 1959, Hughes Aviation purchased a B-25J-30, s/n 44-31489, from Skyways Systems at Miami, Florida. She was registered as N3513G. Hughes owned the B-25 until 1971. Further details are unknown.

N3968C

A second B-25 of Hughes was N3968C, a B-25C-1 with the U.S. s/n 41-13251. This was one of the few early B-25 models that were used in civil aviation. After her USAAF career, she had some Texan owners. Her initial civil registration of NL75635 was assigned in December 1947. In the early 1950s, she even was exported for service with the Dominican Air Force. In 1952, she was sold by the Dominican Air Force to Charles Mathews and Company, Miami, Florida respectively to the Babb Company of Newark, New Jersey.

man of the Board of RKO Pictures, and the Director of Hughes Aircraft. At some point, the airplane was converted into an executive transport for the company. Later, she was taken to the strip where Hughes had his other planes parked. She was grounded in Culver City, which is where she set until acquired by Antelope Valley Air Museum of Lancaster and in July 1974 was transferred to Lancaster.

N75831

A third Mitchell of the Hughes company was one of the oldest B-25s. This was s/n 40-2168, the fourth production model. In 1943, this airplane was modified as General Arnold's personal airplane. In January 1947, after disposal at McClellan Field, California, she was sold to Charles Bates of Chattanooga, Tennessee. In April 1947, she was registered as NL75831. In 1948, she was sold to Bankers Life and Casualty Company from Chicago, Illinois. At the time, she was flying as "El Jarocho" and registered as N75831.

N3968C was converted into an executive transport for the Hughes Tool Company. She was painted overall white with red stripes.
(napoleon130.tripod.com)

form Baker Hughes Incorporated in 1987. In 1985, Hughes Aircraft Company was sold to General Motors. The GM-Hughes Electronics Corporation was established as a wholly owned subsidiary of GM and itself has two subsidiaries: Hughes Aircraft Company and Delco Electronics Corporation.

Her current registration of *N3968C* was assigned in July 1952. The airplane was sold to Hughes Tool Company in January 1953. She was to be the VIP transport aircraft for Mr. Noah Dietrich. He was the right-hand man of Howard Hughes. In fact, Dietrich was the Director-Vice President of the Hughes Tool Company, Director-Chairman of the Board of the Executive Committee of TWA, Chair-

In June 1951, she was sold to Hughes Tool Company. Howard Hughes used her for corporate activities. In June 1962, she was sold to Hughes/Acme Aircraft in Lomita, California. Acme Aircraft Co. was a company started by Charles Roger Keeney in 1945 at California's Torrance Airport, then Lomita Flight Strip. In the past decades, Acme Aircraft had provided maintenance services for many

The Mitchell N75831 seen at Torrance Airport, California in May 1964. At the time, she was owned by Hughes/Acme Aircraft in Lomita, California. (Nick Williams)

famous Southern California aviators, including notables such as Hollywood pilots Paul Mantz and Frank Tallman. The registration of the B-25 was cancelled in May 1965 as she was reported as salvage. In October 1966, she was owned by Aero Service from Wichita Falls, Texas. The registration was changed to the current N2825B in November 1966. She was sold back to Acme Aircraft in January 1967 and was there until August 1971, when she was sold to John Silberman from Savanah, Georgia. In the 1980s, she was still flying as "Proud Mary". Since 2002, she has been operated by the American Airpower Museum in Farmingdale, New York, and flies as "Miss Hap".

L.B. SMITH AIRCRAFT CORPORATION, MIAMI, FLORIDA

A company that did not own a B-25 but did major maintenance and overhaul was L.B. Smith. In the context of this book and the aforementioned maintenance and modification companies, it is nevertheless interesting to pay some attention to this company. In 1947, this company was founded in Miami, Florida, by company president L. B. Smith. It quickly became one of the foremost aircraft conversion, overhaul and modification centres in the United States. Their most famous conversions were made on the Curtiss C-46 Commando and civilian conversions for ex-military Douglas A-26 Invaders and Fairchild C-82 Packets. They also did executive aircraft interiors for many types from the Douglas DC3 to the Lockheed JetStar. L.B. Smith was strongly associated with the similar companies Aerodex Inc. and Aerosmith Inc. They were primarily engaged in overhaul, modification, and sale aircraft and component parts. They were located in hangars, buildings and shops at Miami International Airport. Although separate entities, the companies had the same directors, and L.B. Smith was the chairman of each board of directors and owned most of the stock in each company. L.B. Smith went into liquidation around 1963. Aerodex Inc. was founded in 1946 by W.H. Hart & H.R. Jenks, overhauling WWII surplus aircraft and parts. They were located just west of Miami International Airport. By the 1960's, they employed 6,500 people and played a big role in overhauling engines for the U.S. Air Force in the Vietnam War. They were blamed for supplying defective parts to the military which caused three crashes. They lost the Air Force contracts and had to lay

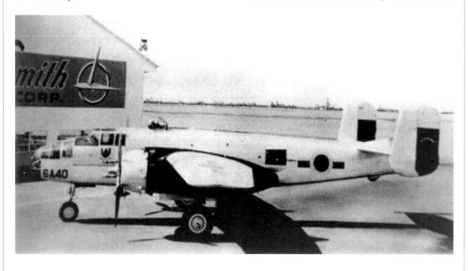

This B-25J #6-A-40 of the Venezuelan Air Force was one of the Venezuelan B-25s overhauled by the L.B. Smith Aircraft Corporation at Miami, Florida. (Dan Hagedorn)

off 3,000 workers. By 1976, Aerodex was declared bankrupt and their assets liquidated at auction.

The L.B. Smith company has overhauled at least 28 B-25s. In 1952, 10 ex-Canadian Air Force B-25s were overhauled by L.B. Smith. They were intended for the Venezuelan Air Force. In 1956/1957, 9 other B-25s of the Venezuelan Air Force were overhauled by L.B. Smith. In December 1963, Venezuela acquired again 9 low-time ex-Canadian Air Force B-25s. Prior to delivery, they were also overhauled in Miami.

A 1977 advertisement from "Trade–A–Plane" of Warbirds of the World Inc. at San Marcos, Texas. Shown in the picture is N9494Z "Laden Maiden".
(Collection Doug Fisher)

JOHN J. STOKES, SAN MARCOS, TEXAS

A great warbird lover was John Stokes. In the late 1970s, he owned eight B-25s. In 1952, John Stokes founded Stokes Construction Co., a building contractor at San Marcos. For nearly 70 years, Stokes Construction Co. has been putting up buildings in Central Texas and elsewhere. The company built some of the highest profile developments in the area. In 1981, his wife Fraye added to the company's resources with years of management, marketing and financial skills benefiting the many business ventures and properties owned and maintained. At its largest, Stokes Construction had as many as 500 employees. In the mid-1970s, John Stokes began to scale back. John Stokes had been active in attracting industry to San Marcos and, in the early 1980s, had served on the City Council and as mayor pro-tem and acting mayor. Stokes was a man of many interests including serving as a colonel in the Confederate Air Force. He owned more than forty warbirds. Among them were eight B-25s. At one time, it was the largest privately owned air force in the world. He used to love to fly those planes. He was a big player in the warbird movement in the 1970s operating as Warbirds of the World Inc. out of San Marcos. Col. John Stokes founded the Central Texas (CenTex) Wing of the Confederate Air Force in 1974, and generously donated time and money to help assure the success of the Confederate Air Force CenTex Wing. In January 2002, the Confederate Air Force was renamed Commemorative Air Force. The Commemorative Air Force is a non-profit organisation dedicated to preserving, in flying condition, a complete collection of the combat aircraft flown by the Allies during World War II. The CenTex Wing of the Commemorative Air Force is located at the San Marcos Municipal Airport, in the only remaining 1943 vintage wooden hangar on the airport.

The eight B-25s owned by Stokes were N2849G, N3337G, N6123C, N9077Z, N9117Z, N9443Z, N9494Z and N9552Z. One of those was a film star and three aircraft were purchased from Dothan Aviation Inc., Dothan, Alabama.

N2849G

This airplane was purchased by John Stokes in 1976 from Cecil Harp, James Orton, David Hileman & Martin David of San Mateo, California where she flew as U.S. Army Air Force #430077 with tail code 9G. But already in April 1977, Stokes sold the B-25 to Richard Sawyer, Clarence, Indiana.

N2849G was purchased by John Stokes in 1976. She flew as U.S. Army Air Force #430077/9G. Here she is shown in 1975. (Collection Wim Nijenhuis)

N3337G

This airplane was purchased by Stokes in January 1976 from J.K. West/R&W Air Ag Inc., Angleton, Texas. In August 1977, Stokes sold her again to Dwight Reimer/Minter Field Air Museum at Shafter, California. In October 1979, she was purchased by Leroy W. Richards, Chico, California. Finally, she ended up at the Castle Air Museum at Atwater, California. There, she was first exhibited as Doolittle Raider 02344 and later as "Lazy Daisy Mae".

N6123C

In November 1965, this B-25 was sold to Aircraft Specialties of Mesa, Arizona. There she flew as tanker #34C but was withdrawn from use and stored derelict at Mesa from 1969 until 1976. She was purchased by John Stokes/CenTex and flown to Chino, California for storage. In May 1977, she was purchased by Kansas City Warbirds of Kansas City, Missouri and restored to airworthy condition.

N9077Z

This was a ship from Dothan Aviation Corp. of Dothan Alabama. She was modified for agricultural spraying and dusting in January 1962. By October 1975, she was sold to John Stokes and in September 1977 was sold again to Charles Skipper, Charles Becker and Jack Jones of San Antonio, Texas.

N9117Z

She flew with Abe Sellards of Safford, Arizona and was operated by Abe's Aerial Service

as tanker #35C. She was sold in November 1965 to Aircraft Specialties in Mesa, Arizona. In 1975, she was sold to John Stokes, doing business as CenTex Aviation in San Marcos. In April of 1977, she was sold to Robert Lumbard and Dennis Kincaid, Rialto, California.

In the late 1970s, after N3337G was sold by Stokes, she flew a short time as "Shady Lady" with a solid nose. (Collection Wim Nijenhuis)

N6123C at Mesa in 1974. She flew as tanker #34C, was withdrawn from use and stored derelict at Mesa from 1969 until 1976 before she was purchased by John Stokes. (eLaReF)

Side view of N9077Z nosed up to John Stokes' CenTex Aviation hangar in San Marcos in 1977. She still has the colours and markings of the Dothan Aviation Corp. (Mike/bluehawk15)

N9443Z

A second ship bought from Dothan Aviation Corp. was N9443Z. She was purchased by Stokes in October 1975. She was sold again in April 1976 to Richard Shepard & William Knight, Ozark, Alabama. In December 1981, she was displayed as "Piece of Cake" at the Heritage of Flight Museum, Springfield, Illinois. Circa 1985, she went to Aero Trader, Chino, and was dismantled and stored at Ocotillo Wells. Her fuselage centre section was used to rebuild N898BW and the cockpit section was used as a studio mock-up for the filming of "Forever Young" and was then rebuilt as a static forward fuselage display at the Discovery Channel Store in Washington, D.C.

N9117Z somewhere in the 1970s. Her colours as tanker #35C of the former owners Abe's Aerial Service and Aircraft Specialties at Mesa have become pretty weathered and nearly disappeared. She was purchased by John Stokes in 1975. (Collection Wim Nijenhuis)

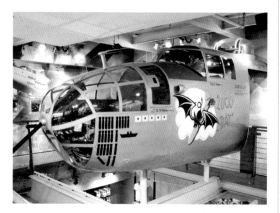

N9443Z at Wheelless Airport in April 1975. The aircraft was operated as a sprayer by Dothan Aviation Corporation. The cockpit section is photographed at the Discovery Channel Store in Washington, D.C. in July 2000. (Geoff Goodall, Wim Nijenhuis)

N9494Z

This airplane was a film star. In December 1968, she was sold to Filmways Inc., Hollywood and flew in the film "Catch-22". After her film career, she was sold to the Confederate Air Force at Harlingen, Texas in February 1970 and to Wayne Turner/Michael Zahn at Lubbock, Texas in February 1972. John Stokes purchased this B-25 in September 1975. He flew her as "Laden Maiden" like she did in the film "Catch-22". In April 1978, the airplane was sold to John Hawke's Visionair International Inc., Miami, Florida and ferried to the U.K. for filming "Hanover Street".

N9552Z

The third B-25 Stokes purchased from Dothan was N9552Z. In September 1960, she was sold to Dothan Aviation and modified as a tanker. She served for many years with Dothan and the company sold the ship in August 1975 to John Stokes. After four years, she was sold to Henry W. Fisher, Donald W. Ericson and Robert E. Thompson at Bloomingdale, Michigan in August 1979.

N9552Z at San Marcos in January 1976. At the time, she was owned by John Stokes. Note the old makings from Dothan are just barely visible. In October 1981, the bomber was donated to the Confederate Air Force.

(John Kerr)

N9494Z "Laden Maiden" of John Stokes on the ramp at Harlingen, Texas in 1977. His name is written below the cockpit together with the name of the pilot Bill Grey. (Geoff Goodall)

TOM REILLY, KISSIMMEE, FLORIDA/DOUGLAS, GEORGIA

Like Aero Trader in Chino, Tom Reilly is one the most famous B-25 restorers. Tom Reilly has been in the warbird restoration business since 1971 after a chance flight in a P-51 Mustang in Kissimmee, Florida. Not long after that flight Reilly had an opportunity to purchase some North American Yale Training airplanes from a friend in Canada. He ended up buying thirteen of them and restored five to flying condition.

The next big project was a derelict B-25 he located in 1977 at Caldwell, New Jersey. This airplane with the civil registration N6578D had served as the camera ship for the film "Battle of Britain" in 1968. She had lain abandoned for years before Reilly found her. He bought the airplane and got her flying again. This was his first B-25 project and it started a long love affair for Tom Reilly with bomber-type aircraft.

Reilly's big break came when he met Harry Doan from Daytona Beach, who graciously allowed Reilly to fly his B-25 for his type rating. Reilly's next big break came when he got a request from Bob Collings to do a pre-buy and purchase on a B-25 in Washington

The workshop of Reilly Aviation at Kissimmee in July 2000. The B-25 tail is from "Killer B". (Wim Nijenhuis)

State. Reilly restored Collings' B-25J "Tonde-layo", and also his B-17G "Nine O Nine" and B-24J "Witchcraft".

In all his years of being in the business, Reilly has worked on at least thirty four major restorations which include at least ten B-25s, three B-17s, a B-24, a P-40, several F4U Corsairs and nine T-6/SNJs. Reilly was located in the Orlando/Kissimmee area of Florida for over 30 years. He had a big hangar with a machine/workshop attached. He and his wife Sue, also had a one-story building that housed offices and a museum of all kinds of aviation stuff, mostly WWII pictures, warbird parts, etc. Opened in 1988, The Flying Tigers Warbird Restoration Museum at Kissimmee Municipal Airport was always one the favourite attractions in Orlando. Amongst the static aircraft on display were aircraft from

the early day's right up to the Vietnam era. The flyable aircraft also included the B-25 "Killer B". Reilly also ran the Tom Reilly's Restoration School where students in a week-long hands-on course learned everything from hydraulics and brakes, control systems, propellers, electrical systems and radios, sheet metal fabrication and welding, wood and fabric techniques and engines and fuel systems. Unfortunately, in August 2004, Hurricane Charley destroyed the restoration shop and museum and Reilly closed in 2005. He moved to Douglas, Georgia, to work part-time for Don Brooks restoring his P-40E. Reilly then helped complete a totally stock restoration of Brooks' PT-17 Stearman.

The latest project that Reilly has found is the North American Aviation XP-82 Twin Mustang prototype. This aircraft is only one of

two that still exist in civilian hands and is being restored in Douglas. Reilly has also provided many artefacts from his Flying Tigers Museum to the WWII Flight Training Museum in Douglas, Georgia. This is a fine little museum related to a WWII Flight Training program located in Douglas. The site is the actual location of the flight training school which operated during WWII. The buildings are original with several being restored and utilised.

Some B-25 projects of Tom Reilly

Tom Reilly has done maintenance and restoration work on B-25 Mitchells like: N898BW, N1042B, N2849G, N3476G, N3969C, N3970C, N5865V, N6578D, N9079Z, N9167Z, N9621C and N62163.

Below a few B-25 projects are described that have been taken over by Tom Reilly in the past.

N898BW

In July 1985, she was trucked to Tom Reilly for restoration. Because her centre section was left at Aero Trader, the centre section of s/n 44-28765 was used for restoration. After restoration, she flew for the first time in 1988. She was finally sold in 1991 to Mr. Gene Rayburn of Arizona who operated her for several years in a RAF colour scheme. Nowadays, she flies as "Axis Nightmare".

N2849G

in December 1978, this airplane was sold to Reyline Aviation at Kissimmee, Florida, and was loaned to the Flying Tigers Air Museum at Kissimmee. She was displayed as

The glass nosed N898BW after restoration with the centre section of s/n 44–28765. She operated for several years in an unusual RAF colour scheme. She is photographed here at Tom Reilly's workshop at Kissimmee in 1992. (Larry Johnson)

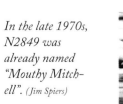

In the late 1970s, N2849 was already named "Mouthy Mitch-ell". (Jim Spiers)

"Mouthy Mitchell". In January 1979, she was sold to Medusa Holding Co., Grand Cayman Island, January 1979. In May 1979, she was sold to Frank Guzman and in October 1980, to the Georgia Historical Aviation Museum, Stone Mountain, Georgia. In 1984, she was reported damaged in landing accident. She went to Tom Reilly and was stored disassembled at Kissimmee. In 1993, a restoration was started, but later she was stored. Parts of the bomber were in the early 21st century used in the restoration of the B-25 of the Pacific Aviation Museum at Honolulu, Hawaii.

N3476G

This former tanker of Earl Dodge from Anchorage, Alaska and Robert Schlaefli of Port Orchard, Washington, was purchased by the Collings Foundation in 1984 and restored by Tom Reilly. She was flown at that time in

Desert Sand colours as "Hoosier Honey".
On 10 June, 1987, she crashed at Minute Man Field in Stowe, Massachusetts and was rebuilt and based in Houston, Texas. In 2002, her name was changed to "Tondelayo".

The Collings Foundation is a non-profit, Educational Foundation. The purpose of the Foundation is to organise and support "living history" events and the preservation, exhibition and interaction of historical artefacts that enable Americans to learn more about their heritage through direct participation. Currently, the B-25 "Tondelayo" is still operated by the Collings Foundation.

N5865V

B-25J-30, s/n 44-30988, was registered as N5865V. In the 1960s, when she was purchased by Air Services Inc. and operated by Aero Industries respectively, both at Addison, Texas, she was used to test electron-

ics including Doppler navigation radar and side-looking radar. She was sold in 1971 to Robert A. Mathews of Jacksonville, North Carolina. The next owner used her as a cargo aircraft between Central America and southern Florida. In 1972, she was reported derelict at Ft. Lauderdale, Florida.

She was sold at a Public Sale in January 1976 after an unsuccessful search for the owner in Panama. She was purchased in 1978 by Tom Reilly and transferred to Kissimmee in 1979.

After restoration, she flew as "Big Ole Brew 'n Little Ole You". In 1988, she was sold to Craig Tims of Roanoke, Texas. She was transferred to the Confederate Air Force, Harlingen, Texas in June 1988. In 1991, she was transferred to the American Airpower Heritage Flying Museum, Midland, Texas and in 1993 assigned to the Southern California Wing of the CAF, Camarillo, California. By

N3476G was restored by Tom Reilly and flew with Desert Sand colours as "Hoosier Honey". (Collection Roger Ritter)

The beautifully restored N3476G now as "Tondelayo" of the Collings Foundation in the colours carried by one of the airplanes of the 345th BG, 500th BS during World War Two.
(Collection John Beach)

N5865V at Fort Lauderdale, Florida in 1974. She still has the civil paintwork from her days as an executive transport fitted with an air stair door and other refinements. In 1977, she was damaged to her tail from another airplane being pushed into her during a hurricane that rushed through South Florida. In 1978, she was bought by Tom Reilly and transferred to Kissimmee in 1979 for restoration. *(Steve Williams)*

July 1983, the restored "Big Ole Brew 'n Little Ole You" at the Arapahoe Air show, Colorado. At right, close up of the nose art. (Collection G. Verver, Zane Adams)

1994, restoration was started by the Southern California Wing and in May 2016, after a 23-year restoration, she became the only known flying PBJ-1J.

N6578D

This was the first B-25 project of Tom Reilly. The airplane was owned by John Hawk and used in the U.K. for the film "Battle of Britain". In January 1969, after completion of the filming, the airplane returned to the U.S.A. She was reported as derelict at Caldwell-Wright Field, New Jersey in 1970. For the next few years, she was locked up in a legal battle until March 1975, when she was sold to Ten Plus One, Inc. They sold her in April 1977 to Tom Reilly. Restoration of the bomber was begun at Caldwell-Wright Field. In February 1979, she was ferried from Caldwell to Florida and the restoration to airworthy status was finished in Kissimmee. In February 1979, the bomber was sold to B-25 Bomber Group Inc., Ocala, Florida. During 1981, restoration to airworthy status was completed and the airplane was in natural aluminium finish with the nose art "Chapter XI" with the 5th Air Force logo and "25th Bomber Group". They sold her in August 1994 to Dan Powell of Boerne, Texas and the bomber was then flown as "Lucky Lady".

The camera ship N6578D as derelict at Caldwell-Wright Field, New Jersey. She was ferried to Kissimmee for restoration, which was completed in 1981. (Collection Wim Nijenhuis)

In 1998, based at Franklin, Virginia, she was withdrawn from use and placed in open storage. Eventually, she degraded to derelict condition. In April 2015, she was purchased by Reevers Warbird Roundup in Australia for restoration. See chapter Australia.

N9079Z
In the early 1970s, this airplane flew as an insect sprayer called "Big Bertha" and in 1974, she was donated to the SST Aviation Museum.

Bottom:
N9079Z "Panchito" of Larry Kelley photographed at Grimes Field, Urbana, Ohio in April 2017. (Dennis Nijenhuis)

The former camera ship of John Hawke was restored by Tom Reilly and named "Chapter XI". Here she is shown in 1986. **Below:** *Aa close up of the nose art.*
(Collection Larry Johnson, John Hevesi)

After the museum closed, they sold her in October 1983 to Pat O'Neil, Robert Bolin and Jack Myer of Wichita Falls, Texas. Tom Reilly was given the task of restoring her to her former glory. He began a total rebuild back to its original "J" model configuration, completing the airplane in 1986 for the new Texas owners. After arriving in Texas, the bomber received the nose art and markings as "Panchito" from the 396th Bomb Squadron, 41st Bomb Group. Her first flight after restoration was in March 1986. She was used in the filming of the movie "Pancho Barnes".

In September 1988, she was damaged when her landing gear collapsed during roll-out. She was subsequently repaired. She was sold again to Richard Korrf of Lewiston, New York in March 1991. By April 1992, she was sold to Aero Classics of Lewiston, New York. In 1997, she was purchased by Larry Kelley, chief-pilot and owner of Rag Wings and Radials, Wilmington, Delaware. She now resides at the Delaware Aviation Museum in Georgetown, Delaware.

N9167Z

In January 1959, this B-25 got her civil registration N9167Z. She changed ownership several times and in 1972, she was sold to Crosby Enterprises Inc., Wauwatosa, Wisconsin. Dr. Crosby used her frequently within his medical practice flying to patients in remote areas. Some of Dr. Crosby's calls took him into rough terrain including some Indian Reservations with small and rugged landing strips. When not in commercial use she was also flown to local air shows. During this time, she was operated as "The Devil Made Me Do It".

In 1987, she was obtained by Tom Reilly and flown to Kissimmee for maintenance. The airframe was reworked and returned to a standard glass nose from the modified nose that was on the airplane. In June 1988, she was sold to Mid South Lumber Company of Cropwell, Alabama. By 1992, she was sold again, and her registration was changed to N345BG. Later, she flew as "Martha Jean" and currently as "Georgie's Gal".

N62163

Tom Reilly not only restored this bomber, but also maintained and operated this Mitchell; then purchased and moved her to her new home base at Douglas, Georgia. In December 1963, this former RCAF B-25J-30 with s/n 44-86697 was obtained by the Fuerza Aerea Venezolana, who operated her until she was retired in the early 1970s. She was stored in Venezuela until 1992. The bomber was sold to Aztec Aviation Consulting of Miami and was purchased in September and auctioned off at the Tico Warbird air show. The winning bid went to Bill and Marie Leary (Tricon Aero Corporation, Elmwood Park, New Jersey) in September 1992.

During the 1970s and 1980s, N9167Z "The Devil Made Me Do It" also flew at local air shows before she was restored by Tom Reilly.
(Eduard Marmet, Paul Thallon)

By the end of the 1980s, Tom Reilly did maintenance work and refitted the standard glass nose on N9167Z. In 1992, she received the civil registration N345BG and was later named "Martha Jean". (Mark Fisher/twinbeech.com)

Reilly was at the air show and introduced himself and told Bill that he specialised in rebuilding B-25s. Tom was hired to breathe life back into this old bomber, and a ground-up rebuild was started in 1993. In February 1993, she was registered as N62163. The bomber was finished in an USAAF North African-style camouflage with markings that included a yellow border around the national insignia signifying Operation Torch. The British-style tail fin flash was applied as well as the nose art "Killer B". The bomber made her first post-restoration flight in August 1995. Since May 2011, the bomber is owned by Tom Reilly Vintage Aircraft Inc., Wilmington, Delaware.

"Killer B", the B-25 of Bill Leary in desert camouflage parked near the museum and workshop of Tom Reilly at Kissimmee in July 2000. (Wim Nijenhuis)

VARIOUS COMPANIES AND OWNERS

AEROCRAFTERS INC., SANTA MONICA, CALIFORNIA

For more than a decade, a B-25 was owned and flown by Aerocrafters in the colours of the well-known "Sunday Punch". The company Aerocrafters was privately held in Santa Rosa, California and was specialised in aircraft servicing and maintenance. It was established in 1992 and incorporated in California. In January 1992, Steve Penning and Lynn Hunt started Aerocrafters Inc., specialising in the restoration of warbird and vintage aircraft with an FAA certified repair station for avionics. Aerocrafters was started as a business and to indulge their passion for military airplanes. They were located at the Santa Rosa Airport and have grown to occupy nearly 50,000 square feet of hangar, shop, and storage space. Airplanes that were restored or worked on at Aerocrafters included fighters, bombers and some jet fighters. The B-25 Mitchell they owned, was housed in a subsidiary called Mitchell Mania, LLC. The owners of Mitchell Mania were Bill Manly, Steve Penning and Lynn Hunt. Aerocrafters evolved into Sonoma Jet Center in 2004 and became a full service FBO at the Sonoma County Airport in Santa Rosa.

Their B-25 was delivered in June 1945 and was flown directly to a storage facility in Independence, Missouri. In December 1951, the airplane was transferred to the Royal Canadian Air Force and was used as a multi-engine trainer until being struck off charge in 1961. She was sold by the RCAF to Hicks and Lawrence, Ltd. of Ostrander, Ontario, Canada. In 1965, she returned to the U.S. and was registered as N543VT. Af-

April 2003, N325N "Sunday Punch" of Aerocrafters at Nut Tree Airport, Vacavill, California. At right, the warbird in 2011 with the same name and additional nose art as carried by the 12th BG during the war. (Bill Larkins. S.R. Breitenstein)

ter she had some different owners, she was sold again in August 1982 to the Canadian company G&M Aircraft of St. Albert, Alberta. There she flew as a B-25 fire bomber, the last in the world. She flew in an overall yellow paint scheme and was withdrawn from service in 1992. Penning and Hunt discovered the B-25 in Canada in 1998 and she was primarily a gunship instead of a bomber. In October 1998, she returned to the U.S. purchased by Mitchell Mania. Her new home became the Charles M Schulz-Sonoma County Airport in California, just north of San Francisco. Their company, Aerocrafters, put the airplane through a three-year restoration, and she eventually emerged as "Sunday Punch", carrying the civil registration N325N. The overhaul included the removal of the fire bomber's retardant tanks and a return to stock Mitchell condition. A major portion of the transformation was the conversion to a solid eight-gun nose, as the type had in World War Two's Pacific Theatre. The USAAF paint scheme depicted on "Sunday Punch" was representative of a B-25J of the 81st Bomb Squadron, 12th Bomb Group. The name "Sunday Punch" was a salute to workers at the K-25 Plant in Oak Ridge, Tennessee, who donated their Sunday overtime pay to raise funds for the airplane. The bare metal fuselage included a distinctive and colourful nose featuring the name, pin-up art and a large, fang baring face on the front surrounding the gun installation. In May 2012, she was sold to Fagen Fighters WWII Museum at Granite Falls, Minnesota. She was restored by Aero Trader, Chino in 2012-2013 and fitted with a glass nose. She is now flown in an Olive Drab and Neutral Grey scheme as "Paper Doll".

It should be noted that Aerocrafters had another B-25. A short time after the above-mentioned B-25 was found in Canada, Lynn Hunt found out about four B-25s the Venezuelan government had abandoned in the jungle. He rescued one of the airplanes from the overgrowth, and she sat in the shop, in sections, waiting to be rebuilt.

ALBERT TROSTEL & SONS COMPANY, MILWAUKEE, WISCONSIN

In the early 1950s, one of the major tanneries of Milwaukee flew with two B-25s. NL75754 and N75755 were two executive airplanes of Albert Trostel & Sons Company. This company was founded in 1858 and was based in Milwaukee, Wisconsin. Albert Trostel & Sons built its business on providing leather boots for Union soldiers in the Civil War, as well as leather accessories for horse-drawn transportation. The company continued to grow, refining the art of leather processing during the industrial age and adapting and diversifying to create leather seals and gaskets for military machinery during World War II. Starting in the 1950s,

Albert Trostel & Sons began further diversifying and acquiring other companies, most notably the acquisition of Eagle Ottawa Leather Company in the early 1960s. The leather operations of Albert Trostel & Sons were merged with Eagle Ottawa, a supplier of automotive leather, and the merged business retained the Eagle Ottawa name. In the 1950s, Albert Trostel & Sons also opened a new entity in Lake Geneva, Wisconsin. Albert Trostel Packings, Ltd. (renamed Trostel, Ltd. in the 1990s) was established to produce leather packings and rubber seals and later expanded to include synthetic rubbers, thermoplastic elastomer

NL75754 of Albert Trostel & Sons Tanners, Milwaukee. The company owned the airplane for six years. **Below:** A 1948 postcard of the aircraft. (Collection Wim Nijenhuis)

and cast urethane products used across various industries. In 1962, following the death of the founder's grandson Albert O. Trostel, Jr., long-time executive Everett G. Smith became president of Albert Trostel & Sons. In the 1960s, separate from his involvement with Albert Trostel & Sons, Everett Smith had taken controlling interests in several other companies, including Maysteel, a sheet metal fabrication company, and Blackhawk Leather, a manufacturer of split leathers for shoes, garments and handbags. Both businesses were located in south-eastern Wisconsin. In 1974, Everett Smith Group (ESG) was formed as a holding company for Mr. Smith's interests in Maysteel, Blackhawk and his growing ownership stake in Albert Trostel & Sons. ESG continued to invest in new industries including the steel casting business, thermoplastics businesses and electronic assembly. In addition, by 2010 ESG had acquired full control of Albert Trostel & Sons and its subsidiaries Eagle Ottawa and Trostel, Ltd. The Trostel tannery was a major Milwaukee employer for nearly 100 years.

Both B-25 airplanes of Albert Trostel & Sons were J-30 models. They had the USAAF serial numbers 45-8882 and 45-8883 and were never delivered to the military. They were completed when the Fairfax plant was closed and were delivered directly to the disposal locations at Walnut Ridge and Altus respectively in October 1945. In August 1948, s/n 45-8882 was purchased by Albert Trostel & Sons and registered as NL75754. The company owned the airplane for six years. In October 1948, she was registered as N32T. The airplane was sold to J.J. Ryan, Arlington, Virginia in September 1954.

The other ship with s/n 45-8883 was sold to Albert Trostel & Sons in July 1948 and registered as N75755. This airplane served with the company for nearly two years and was sold to Northern Pump Company in May 1950.

In October 1948, the civil registration was changed from NL75754 into N32T. Also the paint scheme was changed. (Collection Wim Nijenhuis)

ARTHUR JONES, SLIDELL, LOUISIANA

There is still some misunderstanding about the registration of the B-25s owned by Arthur Jones. He is the inventor and founder of the Nautilus exercise equipment company. Prior to Nautilus, Arthur Jones was an importer of wild animals from South America, had a zoo in Slidell, Louisiana, and a television show called "Wild Cargo". He was the owner of three B-25 Mitchells, which he used to transport his animals. Arthur Jones was perhaps the most all-time controversial entrepreneur of strength training. On his adventurous road to fame and fortune, Jones learned to fly on ramshackle airfields throughout Oklahoma and matured into a fearless barnstorming pilot. Later, he operated airlines in Latin America, flew planes throughout Africa, and owned and operated an import/export enterprise specialised in capturing and transporting snakes and a variety of reptiles and exotic animals. During the 1960s, Jones doubled as a filmmaker. He made a series of TV programs that aired as "Wild Cargo" in the United States. In the mid-1960s, he moved his family to Rhodesia, where they lived for two years until the government took exception to his wild-game business and seized his assets, forcing his return to the United States. Together with the help of his son Gary, he was responsible for the idea, design, and development of Nautilus gym equipment. From 1974 through 1982, Jones dominated the commercial fitness market and claimed more money was spent on Nautilus machines than on the combined sales of all commercial gym equipment purchased. In Lake Helen, Florida, Jones built a $ 75 million corporate headquarters, which incorporated sound stages, film editing rooms and two homes. Following the sale of Nautilus Sports/Medical Industries, Inc., Jones founded the Med X Corporation to pursue research and development projects he began in the 1970s. When he sold Nautilus Sports/Medical Industries in 1986, he owned a large farm located north of Ocala, Florida. The 350 acres of prime real estate housed the largest privately owned airport, including three used Boeing 707 airliners and the largest exclusively owned collection of exotic wild animals. He founded a "fly-in" community in Ocala called Jumbo-lair Aviation Estates, whose most famous resident is actor John Travolta. The three B-25s of Arthur Jones were N3453G, N7947C and N92280.

N3453G

The B-25 with U.S. s/n 44-86844 had registration number N3453G. She was purchased by Arthur Jones in April 1963 from Columbus L. Woods/Woods Body Shop at Lewistown, Montana. According to her FAA file, she was reported to the FAA as being scrapped in 1968, but then was noted as derelict at the New Orleans-Lakefront Airport, Louisiana in 1977 with gutted interior. Further details are unknown.

N7947C

Another B-25 was s/n 44-30129. In June 1958, this bomber was sold to P. J. Murray of Oxnard, California and registered as N7947C. After she had some different owners, she was sold to Arthur Jones in January 1963. She was named "Wild Cargo". The next month, while carrying a shipment of some 2,000 snakes and alligators of a wild animal show, the right engine died just outside Cincinnati. She was unable to operate the landing gear and the co-pilot bailed out and the pilot circled around Lunken Air-

The B-25J N3453G sometime in the mid-1970s at New Orleans-Lakefront Airport, Louisiana. She was not in very good condition and was declared derelict by 1977. (jdanieljr)

On 21 February, 1963 the crippled N7947C "Wild Cargo", loaded with some 2,000 reptiles, made a wheels-up landing at Lunken Airport. Smoke is still coming out of the cockpit. She was eventually auctioned and purchased by Walter Soplata who took the airplane to his secret backyard warplane collection in Newbury, Ohio.
(Cincinnati Aviation Heritage Society & Museum)

port, Cincinnati to reduce fuel. The second engine started to fail as the pilot brought the B-25 in for a gear-up landing. After landing on the belly of the airplane, the airport needed three days to round up most of the snakes. The airplane was dragged off the runway and lifted to again sit on her landing gear by Cincinnati Aircraft, Inc. But Arthur Jones never returned to claim the airplane, so she was eventually auctioned off by the local sheriff's office. Walter Soplata purchased the airplane and took her to his house in Newbury, Ohio.

N92880

The third B-25 was an ex RCAF airplane with s/n 44-30947 purchased by Arthur Jones in April 1963. She was registered as N92880. By 1968, she was withdrawn from use and stored at New Orleans-Lakefront Airport,

Louisiana. Eventually she became derelict. By 1980, she was sold to Hamilton Agricultural Machinery Museum, Petal, Mississippi. Hamilton purchased the airplane for back ramp fees. The Mitchell was disassembled and transported to Petal for reassembly at the Mississippi Machinery Museum. In 1998, she was sold to Big Kahuna's Waterpark and was displayed in an overall yellow colour.

As mentioned above, there is still some misunderstanding about the registration of these airplanes. The airplane s/n 44-30947 was many years incorrectly referred to as 44-86844. Probably this was caused because Arthur Jones owned the B-25s and all were registered with a single FAA N number as he was having trouble registering some of the airplanes. He simply removed all the identification plates from the airplanes and

put the same FAA N number on all the airplanes. Number 44-30947 survived and was falsely represented as 44-86844. The forward fuselage section of 44-30947 is now being used to rebuild the B-25 "Sandbar Mitchell" at the Warbirds of Glory Museum at Brighton, Michigan.

AVIREX, TENAFLY, NEW JERSEY

A special B-25 was owned by Avirex founder Jeff Clyman. This company was a brand for military apparel. Avirex was originally founded by Jeff Clyman in 1975. Contrary to popular belief, Avirex was not a U.S. military supplier during World War II. By 1978, the brand flourished with "The Cockpit" catalogue. The U.S. Air Force began to purchase A-2 flight jackets, which led to them becoming an official government supplier in 1987. This was followed by more government agencies, U.S. Navy, U.S. Army, NASA, and many international armed forces. In 2006, Jeff Clyman sold the Avirex brand and trademark as well as the Avirex sportswear and Hip-hop fashion lines in Europe, Japan, and the United States. The new owner of the Avirex brand in the United States was Eckō Unlimited. Avirex Ltd., now Cockpit® USA, has remained a name that is known for its craftsmanship, quality, and authenticity. The company is recognized worldwide as one of the leaders in fashion and function for military bomber jackets.

The B-25 of Avirex was one of a handful of early production examples which, when originally built, looked different from all subsequent B-25s, having straight wings with a constant dihedral all the way to the wing tips. During World War II, the airplane was rebuilt to later B-25 configuration and modified for use as a staff transport for General Henry "Hap" Arnold. After the war, she was registered as N75831 and used by Howard Hughes in the 1950s. As N2825B, she served as an executive transport for various companies until she was restored in the 1970s, still in transport configuration. In December 1983, her modified solid nose was replaced with a standard greenhouse nose when she was owned by Fighting Air Command, Dallas, Texas. By November

N2825B just after she was purchased by Jeff Clyman. She is here at Chino Airport in December 1989. (Collection Wim Nijenhuis)

N2825B in 1995. At the time, she was named "The Avirex Express" on the right side of the fuselage nose after the company name of Jeff Clyman. She was operated by the American Airpower Museum. (Mike Henniger)

1989, she was sold to Jeff Clyman and donated to the American Airpower Museum in Farmingdale, New York and she flew for a time as "The Avirex Express" and now for a long time as "Miss Hap". She has been flying to air shows throughout the country now for over 30 years.

BENDIX AVIATION CORPORATION, LOS ANGELES, CALIFORNIA

The Bendix Aviation Corporation was a manufacturer of airplane parts based from 1929 to 1960 in Los Angeles, California. It was renamed to Bendix Corporation in 1960. It had various divisions spread across the United States. Bendix Corporation, now part of Knorr-Bremse, began with Vincent Hugo Bendix from Moline, Illinois. He invented an automotive electric starting motor drive, called the "Bendix Drive" in 1914 and became a millionaire. The Bendix Drive and the 4-wheel brake system he invented were the basis for the formation of a new company in 1924. The name became Bendix Aviation in 1929 when production began to include

aviation products. In 1931, Bendix sponsored the transcontinental, point-to-point race and began as part of the National Air Races. The Bendix Trophy is a U.S. aeronautical racing trophy and the last Bendix Trophy Race was flown in 1962. During the war, the Bendix Aviation Corporation made aviation equipment, including its brands Eclipse, Stromberg, Pioneer, Scintilla, and Friez. It covers instruments, landing gear, carburettors, magnetos, hydraulics, armament, turrets, and radio equipment. In 1960, Bendix Aviation changed its name to Bendix Corporation to include its other activities, such as automobiles, space, missiles, and energy controls. In the decades between 1970 and 1990, Bendix went through a series of mergers, sales and changes with partners or buyers. Bendix became a Honeywell brand, including the Bendix/King brand of avionics and the Bendix line of brake shoes, pads and other vacuum or hydraulic subsystems. In 2002, Knorr-Bremse took over the commercial vehicle brake business from Honeywell and Bendix Commercial Vehicle Systems became a subsidiary of Knorr-Bremse AG.

During the 1950s and 1960s, the Bendix corporation used several airplanes in testing the avionics and radar equipment being developed by the company. Also, some of the airplanes were used for executive and personnel transport duty. These were corporate owned (or on loan from the military) and assigned to the various divisions as the need arose. These airplanes included three B-25s with the civil registrations N3184G, N5548N and N69345. At the Radio Division, development work was being done on both ground-based radars such as the

General Henry Arnold's private transport s/n 44-28945. After the war, she was registered as N3184G and was used as a passenger transport airplane by Bendix.
(Collection Wim Nijenhuis),

Two times N5548N from the Bendix Corporation in different colours. The second red colour scheme was taken at South Bend, Indianapolis in December 1964. The first blue colour scheme was seen here at Friendship International Airport in the early-1960's. The airplane was used for many years by Bendix as a test aircraft. (Vince Reynolds, Bob Burns)

AN/FPS-3 and MPN-5, airport surveillance radars such as the ASR-3, and also the first airborne weather radar, the RDR-1. This airborne weather radar would become standard equipment on many commercial and civil aircraft. The aircraft were used to test these radars. Starting in the late 1950s, Bendix developed several phased array radars. These radars utilised multiple transmitters and receivers in an array of antennas to allow various electronically scanned antenna beam shapes. To test these transmitted and received patterns it was necessary to fly an

airplane through the beams. The airplane of choice was a B-25 which had room enough for a transmitter to "talk to" the receive array and also a receiver and recorder system to "listen to" the transmit array. The airplane was directed and observed by people using a very precise theodolite and in this manner the antenna patterns were tested and verified. This program was carried out in 1963-1964, and at that time Bendix Aviation Corporation had three B-25s. The numbers N5548N and N69345 were used in test programmes.

N3184G

In 1944, this was General Henry H. "Hap" Arnold's private transport airplane. The bomber was heavily modified with the nose fully faired over and a series of antennas mounted underneath the forward fuselage. In 1960, she was registered as N3184G. In October 1963, she was purchased by the Bendix Corporation, Baltimore, Maryland. Because the aircraft had been modified to carry passengers, she was used for this purpose by the Radio Division. In 1968, she was sold to Santiago Perez/Double A Leasing,

Miami, Florida. In 1972, the airplane was exported to Bolivia and finally crashed in 1976.

N5548N

In May 1946, this B-25H was released to the Oklahoma Agricultural and Mechanical College in Stillwater, Oklahoma for use in their technical program. The Federal Security Administrator took control of the airplane in June 1951 and subsequently transferred her to the Bendix Aviation Corporation in Detroit, Michigan. Bendix used her for many years as a test airplane. She was used primarily to test landing gear and auto pilot systems for the military. In November 1955, she was fitted with a fully operating F101 nose gear for testing. For the Bendix test programmes, N5548N was assigned to the Radio Division of Bendix. In February 1967, she was sold to Barber's Flying Service, Pontiac, Michigan and transferred to Union National Bank of Chicago in September 1970. Finally, in 1992, after a restoration of ten years this bomber was restored with the gun nose with 75mm cannon and is currently flying as "Barbie III".

N69345

This B-25J, now flying as "Betty's Dream" was sold in February 1946 to Lodwick Aircraft Industries of Lakeland, Florida. In June 1946, she was sold to the Bendix Aviation Corporation in Teterboro, New Jersey. In 1948, her civil registration was assigned as N69345. She was used for radar and avionics testing, among which the Bendix PB-10 autopilot. This analogue electronic autopilot-maintained altitude and airspeed by automatic control of the elevators and throttles. Amplifiers drove aileron, rudder, and elevator servo motors. In November 1967, she went to Bendix Field Engineering Corp., Columbia, Maryland. In August 1972, she was sold to Ernest H. Koons, Edmonton, Alberta and registered as CF-DKU. Later that year, she was sold again to Aurora Aviation Ltd., Edmonton, Alberta.

The ice covered N69345 was used to test flight instruments, among which the Bendix PB-10 autopilot. These pictures were taken in the winter of 1947 at Teterboro Airport, New Jersey.

(Collection Wim Nijenhuis)

CLAIRE AVIATION INC., PHILADELPHIA, PENNSYLVANIA

Tom Duffy oversees all cases handled by the attorneys of Duffy + Partners, which include medical malpractice, birth injury, product liability, motor vehicle/trucking accidents, aviation disasters, construction accidents, premises liability and general personal injury. Duffy founded the firm more than 25 years ago, when he left a promising career at a large Philadelphia firm to establish his own practice. A practice that has developed into a very experienced and respected group of partners and staff. Outside of the office, Duffy is a licensed and skilled commercial pilot who participates annually in air shows all over the country. Claire Aviation Inc. was established in 2009 and incorporated in Pennsylvania. Tom Duffy of Claire Aviation owns eight airplanes from the World War II era, including a B-25. He keeps most of his planes in Millville, New Jersey and has another hangar at Red Lion Airport in Burlington County, New Jersey. In 2009, he took part in a special ceremony to open the Indianapolis 500, soaring over the

In November 1967, N69345 went to Bendix Field Engineering Corp., Columbia, Maryland. During her career at Bendix, she was modified. Note the difference of the glass nose with the picture from 1967. She also has a canopy installed behind the cockpit.

(Collection Wim Nijenhuis, Bob Burns)

speedway in his B-25 during the singing of the "Star Spangled Banner" as the crowd of spectators roared.

The B-25 of Tom Duffy is N3155G. She had a long-term restoration with Aero Trader in the late 1970s and again in the late 1980s. She flew as "Bronco Bustin' Bomber" and "Can Do". In the 1990s, she was operated at the Polar Air Museum as "Buck U". In February 2005, after the museum was closed, the aircraft was sold to Historic Aeroplane Works/Warbird Digest at Huntington, Indiana, the publisher of a short-lived warbird magazine. She acquired a green dragon painted on the nose, reminiscent of artwork on some wartime B-25 gunships of the 38th Bomb Group. By December 2006, she was acquired by collector Tom Duffy, who restored the airplane and chose a generic aluminium paint scheme with the inauthentic, but plausible, nose art and name "Take-Off Time." Duffy's B-25 now appears at airshows on the U.S. east coast.

2005, an attractive paint scheme worn by N3155G owned by Tim Savage of Warbird Digest. The B-25 was officially registered to Historic Aeroplane Works, Inc., of Huntington, Indiana. In 2006, the paint was removed, and she was overall natural aluminium finished but with the same green dragon. (Parr Yonemoto, Wayne Dippold)

Currently, N3155G is flown by Tom Duffy of Claire Aviation as "Take-off Time". She is still aluminium finished but with a new nose art and blue colour accents as can be seen in this picture taken in June 2015 at the Mid Atlantic Air Museum WWII Weekend. (Majd Bostani)

The nose art was made by Gary Velasco and is based on a design of pin-up artist K. O. Munson. On the picture in the centre, Velasco is detailing the face a bit more.

(Surfsupusa, fightingcolors.com

DORIS DUKE, NEW YORK, NEW YORK

A Mitchell in hands of a very wealthy owner was N75150. This airplane was purchased by Doris Duke in 1948. Tobacco heiress Doris Duke was the only child of American tobacco baron, James Duke.

When she was born, the press called her "the richest little girl in the world," but Duke grew to be the most reluctant of celebrities. For over 50 years, she avoided publicity. When she died in 1993, her billion-dollar legacy was left in the sole control of her butler. In 1945, Duke became a foreign correspondent for the International News Service,

where she reported from various cities in war-torn Europe. After World War II, she continued her short-lived writing career in Paris, where she worked for "Harper's Bazaar". Her millions, and the elegant life they made possible, seemed to be of great importance to her two husbands, James H. R. Cromwell, an American sportsman and advertising man, and Porfirio Rubirosa, a Dominican playboy and diplomat. She divorced them both. While in Paris, she met and married Porfirio Rubirosa. He was a diplomat, race car driver, soldier, and polo player. He was

an adherent of the dictator Rafael Trujillo and was also rumoured to be a political assassin under his regime. Rubirosa made his mark as an international playboy for his jet setting lifestyle and his legendary sexual prowess with women. Among his spouses were two of the richest women in the world, Doris Duke and Barbara Hutton. Because the wealth of Doris Duke was so vast, the U.S. government drew up Duke's prenuptial agreement. When they presented Rubirosa with the document, he fainted upon the realisation of her net worth. Because of her

In the mid–1950s, N10V was flown by Porfirio Rubirosa, after his marriage with Barbara Hutton. (Collection Wim Nijenhuis).

Right: *Rubirosa in front of N10V at Teterboro Airport, New Jersey before flying to Paris to meet Zsa Zsa Gabor, in 1954.* (Bettmann/CORBIS)

great wealth, Duke's marriage to Rubirosa attracted the attention of the U.S. State Department, which cautioned her against using her money to promote political agenda. Further, there was concern that in the event of her death, a foreign government could gain too much leverage. Therefore, Rubirosa had to sign a prenuptial agreement. After their wedding in 1947, she gave him a B-25 bomber to be refitted for his personal use, a string of polo ponies, and a Ferrari for him to drive at Le Mans. Their union lasted only a year, and Doris Duke never married again. As part of their divorce settlement, he received a 200-year-old house in Paris. Doris Duke was also the president of the Doris Duke Foundation, whose areas of interest have ranged from social and health services to cultural programs.

Zsa Zsa Gabor and Porfirio Rubirosa with his private airplane N10V at Nice Airport in 1955. (Edward Quinn)

N75150

The airplane of Rubirosa was a B-25J with s/n 45-8824 and converted into an executive airplane that he called "La Gansa". He inherited the airplane when he divorced Doris Duke in October 1948. "La Gansa" had the civil registration number N75150. This airplane was purchased by Doris Duke in January 1948 from Frederick Weicker in New York. The ship operated for American Airlines on high altitude weather flying over the U.S., Canada, and Mexico in 1946-1947 and was modified with additional seats, baggage compartment and APU installed. In March 1948, a 320-gallon fuel tank was fitted in the bomb bay. The airplane was destroyed when crash landed at the Trenton Airport in New Jersey on 9 September, 1948.

N10V

Another B-25 that was flown by Rubirosa was N10V. In March 1954, this airplane was sold to Alfred Merhige at Hollywood, Florida. She was an executive transport and operated by Rubirosa. He flew the ship after his marriage with Barbara Hutton. Barbara Hutton was the Woolworth heiress who inherited a fortune at the age of five after her mother committed suicide. She married seven times and her fifth was with Rubirosa. Her marriage with Rubirosa just lasted 53 days. While Rubirosa was married to Hutton, he was chasing the Hungarian-born American socialite and actress, Zsa Zsa Gabor. Hutton gave him a coffee plantation in the Dominican Republic, the B-25 N10V, polo ponies, jewellery, and a reported $2.5 million as a settlement. Rubirosa flew the ship transoceanic in August 1954 to Cannes and again to Nice in 1955. On both trips Gabor was his "passenger". When Rubirosa would not break off his desire for Gabor, Hutton had the B-25 sold. In December 1954, the B-25 was sold to J. O. Willet at Monroe, Louisiana and in October 1955, she was sold to Husky Oil Company at Cody, Wyoming which operated her until 1960. Further described under Husky Oil Company.

HUSKY OIL COMPANY, CODY, WYOMING

The B-25 N10V was a general's plane in the film "Catch-22". She had fold-down steps on the left side behind the waist gunners' position which is not a standard B-25 construction. The air stair door was added as part of the executive modifications performed around December 1955 when she was then owned by Husky Oil/Canam Company of Cody, Wyoming. Glenn E. Nielson founded Husky Oil on 1 January, 1938 in Cody, Wyoming. He was a rancher, a farmer, an oil executive, and a life-long churchman. In the fall of 1937, Nielson and his family moved to Cody where he, along with some partners, acquired the Park Refining Company and commenced business on 1 January, 1938. The company changed its name to Husky at that time and grew into a corporation that in 1978 employed 2,900 people, produced 50,000 barrels per day of oil, and had 60 million cubic feet a day of natural gas production. In 1946, Husky moved to Canada. It went public in 1949, and through some corporate restructuring in 1960, it became a Canadian Company headquartered in Calgary, Alberta, Canada. Husky had five niche refineries (primarily asphalt facilities) located in Cody, Cheyenne, Salt Lake, Lloydminster, Albert, and Prince George, British Columbia, which provided gasoline and oil products to some 1,200 retail outlets across Western Canada and the U.S. However, Husky's specialty was asphalt, and Husky was a significant factor in this business in both

Metamorphose of an airplane. The B-25H still with the cannon nose in its early days with the civil registration NL90399. Executive modifications were completed in December 1955. The airplane was fitted with an air stair door and a pointed tail cone presumably to improve cruise performance.

Below: *The aircraft in August of 1956 showing the paint scheme of the Husky Oil Company and the registration N10V.* (Collection Wim Nijenhuis, Earl Holmquist)

In March 1960, the airplane was sold to Cherokee Flying Service and the name below the cockpit was changed to "Cherokee II".

(Earl Holmquist)

Western Canada, as well as Western United States. Glenn Nielson was known as "Mr. Asphalt" in both the U.S. and Canada. In addition, Husky had some 13 steel warehousing and fabricating plants from Chicago to the West Coast and eight charcoal plants in the U.S. Just prior to this time, Husky had also been a major offshore drilling contractor. An unfriendly takeover in 1978, caused the Nielson's to sell their interest in Husky the following year. After this, Glenn Nielson busied himself in cattle ranching, family affairs, and investments such as cattle feed yards.

The airplane used by Husky Oil was a B-25H-5, s/n 43-4432. After the war, she was transferred to Altus, Oklahoma in November 1945. In June 1947, she was sold to Joe Zeppa of Delta Drilling Company, Dallas, Texas and registered as NL90399. One year later she was registers as N90399. After that, she got several owners and in February 1952, she was sold to Mechanical Productions Inc., Jackson, Michigan and registered as N10V. Her cannon nose was replaced with the bigger eight-gun nose and the wings were fitted with low-drag wing tips. In 1954, she was sold again and in October 1955, she was purchased by Husky Oil Company. There she was further extensively modified and got a pointed tail cone. In January 1956, the airplane was transferred to Canam Company, Cody, Wyoming. After four years, the ship was sold to Cherokee Flying Service in March 1960 and used primarily as an executive transport as Cherokee II. In the late 1960s, she was purchased by Tallmantz Aviation and used in the film "Catch-22" as a VIP-transport. Currently, she is a static display in the Eagle Hangar at the EAA Air Venture Museum in Oshkosh, Wisconsin.

INTER AMERICAN MINERALS CORP., NEW YORK

In the late 1950s, the B-25J-30 with s/n 44-31280 was operated by Javelin Mines Ltd. In May 1945, this airplane was delivered to the U.S. Navy as PBJ-1J. In September 1947, she was sold by the War Assets Administration (WAA) at Atlanta, Georgia to F.K. Trimmer at Charlotte, North Carolina. The same month, she was sold to Whitehead's Air Service at Charlotte and registered as N52998. After that, she was sold a few times and in February 1956, she was registered as N102J when she was owned by the Forest Protection Division of Olin Mathieson Chemical Corp. at West Monroe, Louisiana. She was sold in October 1958 to Inter American Minerals Corp., New York. The airplane was operated by Javelin Mines Ltd. Finally, she was sold to DuPont Aviation, Miami, Florida and Burnelli Avionics Corp., Washington DC. Further records are unknown.

In the late 1950s, this B-25 was operated by Javelin Mines. In the pictures she has both registrations N52998 and the later N102J.

(Collection Wim Nijenhuis)

John Ward Aviation, Coulterville, California

"Old Glory" of John Ward has had a long career over its lifetime, ranging from US-AAF bomber in World War II in the Mediterranean, to civilian fire bomber and tanker. She suffered a forced landing and being placed in storage, completely restored, and finally went back in the air in its current military configuration. The aircraft with s/n 44-28938 was originally delivered in 1944 and served with the 12th Air Force in Italy before returning to the U.S. in 1945. After initially being stored in Texas, the machine was put back into service in 1947, operating in an administrative role at McClellan Field, California. After a decade at McClellan, the machine was declared surplus and sold. She was registered as N7946C. The following years she passed through the hands of numerous owners, spending time as a fire bomber before being restored to its military configuration and wearing the names "Dream Lover" and "Spirit of Tulsa" before being renamed "Old Glory" in 1995. In 2003, the beautifully polished, natural finish B-25 was sold to John Ward Aviation Co. Inc. in Coulterville, California. J.L. Ward Aviation Co. Inc. is a little privately held company in Coulterville, California. It was incorporated in 2003. John Ward has a deep respect for the men, women, and machines of WWII. His father flew as a gunner/engineer in the Aleutians during the war. John Ward kept "Old Glory" flying until August 2019, when she was purchased by David Prescott of Proair Aviation and was moved to her new home in New York. She was purchased to be included in The Hangar at 743's collection. The Hangar at 743 is a unique facility locat-ed at the end of the runway of the airport in Albany, New York. Their private collection is available for tours, venue rentals, and home to the Warbird Factory which offers WWII airplane repair and restoration.

Unfortunately, the plane crashed on the evening of 19 September, 2020 in Stockton, California. She tried to land in an open field and was found to have hit an ir-rigation ditch. Two of the three crew were injured and transported to a local hospital. The aircraft sustained significant damage.

A magnificent picture of a magnificent airplane. "Old Glory" in 2003, just after she was sold to John Ward. (Ernie Viskupic)

NATIONAL MOTOR BEARING CO., REDWOOD CITY, CALIFORNIA

One of the early owners of a civil B-25 was National Motor Bearing. This company was originally founded in 1920 in San Francisco, by Lloyd A. Johnson. He initially established a plant in Oakland, which was moved to Redwood City in 1942. Johnson, a man who had built his company from scratch, went on to invent and patent in 1936 the process

was not accepted by the USAF. She was delivered new to the Reconstruction Finance Corp. in October 1945. She was sold to National Motor Bearing Co. in 1948 and registered as NL66548. The airplane was converted into an executive transport and flown as "The Flying Seal". Further details are unknown.

The B-25 of National Motor Bearing Company at San Francisco in March 1948. She was registered NL66548 and named "The Flying Seal". (Bill Larkins)

of making laminated shims. Shims are thin pieces of metal or composite used to fill in space between components for adjustment of fit in a mechanical assembly. Shims are usually applied to rotating shafts and sliding surfaces where wear or crushing forces affect a component part. The company produced shims and oil seals for transportation, trains, airplanes, automobiles, and ships. It was a key defence industry during World War Two and one of the major employers of the city at that time. It also had two subsidiaries: The Arrowhead Rubber Co., and National Seal Co. In 1956, the company merged with Federal-Mogul Bower of Detroit which propelled National Motor Bearing as one of the top 300 companies of the country. The National Motor Bearing plant in Redwood City eventually closed in 1971, however Federal Mogul continues to operate today.

The B-25 of National Motor Bearing was a B-25J-25 with s/n 45-8829. She was one of the last B-25s produced at Kansas City and

NORTH AMERICAN AVIATION, INGLEWOOD, CALIFORNIA

North American Aviation Inc. was the designer and producer of the B-25 Mitchell. The company with manufacturing plants at Inglewood, California, Kansas City, Kansas and Dallas, Texas produced nearly 10,000 B-25s during the war. The full story of the company and the production of the Mitchell is described in my previous book "B-25 Factory Times". After the war, North American decided to take a second shot at the executive transport market. Between 1942 and 1950, North American Aviation modified six B-25s to RB-25, mostly used for military transport of VIPs and generals. The last and final modified B-25 was the most extensively modified airplane. This B-25J-30 with s/n 44-30975 was delivered to U.S. Navy as PBJ-1J in April 1945. She was transferred to the Reconstruction Finance Corporation in

November 1945 and purchased by Sherman Machine & Iron Works, Oklahoma City, Oklahoma in November 1946. In April 1948, she was sold to Bankers Life & Casualty Co., Chicago, Illinois and registered as NL5126N. In July 1949, North American obtained the surplus B-25. At that time, North American Aviation thought there might be a real market for converted B-25 executive transports and undertook a program to modify standard B-25s into transports. The front section of the fuselage was completely redesigned. She had a longer and wider nose section, making room for four more seats in the cockpit section, and additional windows. Four additional seats were added in the waist section. The bomb bay was permanently closed and replaced with a baggage door and an electric lift at the front of

The "Bulbous Nosed B-25" with civil register number N5126N. She was the most extensively modified B-25, with a completely redesigned front section of the fuselage, additional windows, four more seats, a long tail cone fairing and semi collector rings instead of the "S" type exhaust stacks of the top cylinders. (North American Aviation)

the former bomb bay. Because of the wider nose, a new windshield was needed and the windshield from the Convair 240 worked perfectly. To reduce noise, a circular exhaust collector was used on the engines instead of having several separate exhaust stacks. The airplane had a plush interior, was registered as N5126N, and was unofficially called "Bulbous Nosed B-25". She flew for the first time on 15 February 1950. Unfortunately, on 25 March, 1950, the airplane crashed in a storm at Chandler, Arizona, and killed all seven men aboard. It was the end of the B-25 transport versions.

"The Bloody Nose" of Northern Pump Company in front of their hangar at Northern Pump Airfield at Fridley.

(Canadian Warplane Heritage Museum, Chris Sorenson)

NORTHERN PUMP COMPANY, MINNEAPOLIS, MINNESOTA

Northern Pump Company was created in 1929 from the merger of two Minnesota companies, and held federal contracts with the United States Navy from 1932 through World War II. The Northern Pump Company produced naval guns during the war. The company was busily expanding from 750 workers to more than 11,000 at the height of World War II. Northern Pump Company was created from the merger of two Minneapolis businesses: Northern Fire Apparatus Company and the Pagel Pump Company. In January 1941, Northern Pump left its Central Avenue location for a new plant in Fridley, Minnesota, and in June 1942, created the subsidiary Northern Ordnance Incorporated to fulfil its federal production contracts. John Blackstock "Jack" Hawley, Jr. purchased and managed Northern Pump Company. He secured Northern Pump's first order for the Navy in 1932, which proved a profitable relationship in coming years. In 1950, the Northern Pump Airfield was built at the Fridley location with a hangar for the Northern Pump Company. The airfield was mainly used for the company's B-25. Post-war, Northern Ordnance continued its

naval production until 1964 when Hawley sold Northern Ordnance to Food Machinery Corporation of San Jose, California. Though he sold Northern Ordnance, Hawley retained ownership of Northern Pump and presumably ran it until his death in 1980. Food Machinery Corporation operated the site until 1994, when a series of corporate mergers and sales began. As of 2009, BAE Systems, a defence, security, and aerospace company, owned and operated the Fridley production site.

The Northern Pump Company owned one B-25. This was the B-25J-35 s/n 45-8883, nowadays flying as "Hot Gen!" with the Canadian Warplane Heritage Museum of Hamilton, Ontario. She was never delivered to the military. She was complete when the North American plant at Kansas City was closed and was delivered directly to the disposal location at Altus, Oklahoma in October 1945. She was initially purchased by A. B. Fitzgerald in September 1946. Her registration was assigned as NL75755. In July 1948, she was sold to Albert Trostel & Sons of Milwaukee, Wisconsin and registered as N75755. In May 1950, she was sold to Northern Pump Company. There she was operated as a business airplane. At the time, the B-25 was named "The Bloody Nose" because it was painted red. Northern Pump Company operated their B-25 until 1958. They sold her in November of 1958 to Peter Volid of Chicago, Illinois.

Oil Tool Corp., Long Beach, California

The B-25J-25-NC, s/n 44-30606, now known as "Tootsie" was for a few years owned by the Oil Tool Corporation at Long Beach, California. She was delivered in January 1945 to the USAF for service in different units. In October 1957, she was flown to storage at Davis Monthan AFB. In January 1958, she was sold to C. M. Jasper of Oil Tool Corp. of Long Beach, California and registered as N5249V. In May 1958, her civil registration was changed to N201L. The ownership went to F. E. Fairfield of Oil Tool Corp. She flew in executive scheme as "N2OIL". The Oil Tool Corp. was a small corporation employing about 50 employees. It has been located at Long Beach since 1926. Owner of the company was F. E. Fairfield; vice president and general manager was R. J. Osburn. After two years in August 1960, the B-25 was sold to Baker Aircraft Sales at Long Beach. They sold her in December of 1960 to John F. Dumm & Jere A. Martlin from Anaheim, California. In November 1962, she was sold to the Union Bank in Los Angeles. She was again sold in April 1965 to B. L. Crawley of Cucamonga, California. After that she had several owners and flew later as a warbird named "Tootsie". As of December 2012, "Tootsie" is privately owned by a real estate company called TSM Enterprises of Carson City, Nevada.

The B-25 of Radio Station KOB. On her nose she has the titles "KSTP News" painted over the faded USAF letters.
(Collection Wim Nijenhuis)

A picture from May 1964, taken at Brackett Field, LaVerne, California. It shows N201L as an executive transport with curtains visible inside the fuselage gunners' positions and other windows. She flew in executive scheme as N2OIL.
(sledge39)

Radio Station KOB, Albuquerque, New Mexico

On 5 April, 1922, Radio Station KOB began regular operation as KOB. In 1933, the station moved to Albuquerque, and was later bought by the Albuquerque Journal. KOB was an NBC Red Network affiliate, carrying its schedule of dramas, comedies, news, sports, soap operas, game shows and big band broadcasts during the "Golden Age of Radio". In 1948, Tom Pepperday, owner and publisher of the Journal, signed on Channel 4 KOB-TV, the first television station between the Mississippi River and the West Coast. The stations were acquired by Time Life in 1952. In 1957, they were sold to Hubbard Broadcasting Inc. Hubbard Broadcast-ing sold KOB-AM-FM in 1986. At that time, stations could not share call signs if they were not co-owned. With Hubbard keeping the TV station as KOB-TV, new owner Southwest Radio had to find new call letters for the radio stations. To trade on the well-known KOB identity, Southwest Radio simply added an extra "K" to the radio stations' call letters: KKOB and KKOB-FM. The change took place on 28 October.

Nowadays, KKOB is a commercial AM radio station licensed to Albuquerque, New Mexico. KKOB is owned by Cumulus Media and is the oldest and among the most powerful AM radio stations in New Mexico. Radio Station KOB owned for only a short time a B-25 bomber. After storage at Davis Monthan, the B-25J-35 with s/n 45-8851 was purchased by Radio Station KOB in April 1959. She was registered as N8093H. At the time she flew in polished metal finish, with "KSTP News" titles on her nose. In December 1959, she already was sold again to Airline Trading, Fort Lauderdale, Florida. In September 1960, she was sold to B. K. Jaquith at Fort Lauderdale. In April 1961, her registration changed into N190V. Finally, she was sold to Imports del Sur Inc., Miami, Florida in June 1961 and reported derelict at Ocala, Florida in 1968.

D. RICHARD LAMBERT, PLAINFIELD, ILLINOIS

N5548N stored at Lambert Field in 1980. She still has her red and white colours of the former owner Bendix Aviation Corporation in Detroit, Michigan. (Glenn Chatfield)

D. Richard Lambert was a farmer and aviation enthusiast. Beginning in 1968, his 80-acre farm was converted to a private airfield, known as Lambert Field. At one time, the airfield housed approximately 70 airplanes, owned by private pilots of all ages and experiences. Lambert had learned to fly and over the course of many years owned or co-owned several airplanes, including one B-25 bomber. This B-25H-1 model, s/n 43-4106 with civil registration N5548N, was purchased at an auction in 1969 and Lambert's goal was to restore the warbird and take it to air shows around the Midwest. After Bendix Aviation sold the B-25 off, she went through several owners. She eventually ended up on the auction block in 1969, where she was purchased by Lambert who placed the winning bid of $3,500.

The aircraft was flown from Michigan into the airfield in the Spring of 1970 and a crowd of several hundred people came to watch the airplane arrive. Several times each year, the B-25 was taken up for a short flight. In between, Lambert often taxied her around the airfield. During many weekends in the summer, several hundred people spent their weekends at the airfield. Because of the proximity of several other airports, Lambert Airfield was forced by the FAA to reduce the number of airplanes allowed on the field in 1973. The last time the B-25 was being taxied was in 1977. After six years of unsuccessful attempts to relocate the family-run airport to another site, Lambert tried several new ventures while retaining the reduced airfield operation. In the early 1980s, he decided to sell the B-25. From that time, the airplane was left unattended at the Plainfield airfield. Weary Warriors Squadron of Chicago, Illinois, a group of aviation enthusiasts bought the airplane. After two engine failures, they were able to finally fly the bomber off the farm in 1981. From there, the group embarked on a full restoration of the aircraft. After a 10-year restoration, the Mitchell operated as "Barbie III" and currently is the world's last airworthy B-25H.

RUSS & DON NEWMAN / OLD GLORY INC., TULSA, OKLAHOMA

Old Glory Inc. was a Tulsa father-son team that had restored a B-25 in the 1990s. The Federal Aviation Administration had approved a request from Russ and Don Newman to take passengers up in their restored B-25. The proceeds would directly benefit the Tulsa Air and Space Centre at Tulsa's International Airport. Russ Newman, who ran a Tulsa-based charter flying service, had long been interested in WWII flying machines. Before he had acquired the B-25, he had restored a T-28. The B-25 eventually used a hangar in the Tulsa Air and Space Centre. The B-25 acquired by Newman was N7946C, the B-25J-15 with s/n 44-28938. She had survived the war and a series of deployments at air bases across the country until 1958. The B-25 was converted in 1959 into a firebomber and used by fire services in Washington and Alaska. In 1975, she returned to the U.S. and was sold to Russell DeFrancesco and John Cahill of Cardiff, California. In October 1978 she was sold to Max Power, Inc. of Carlsbad, California and

Jan Ricketts with her artwork of "Dream Lover". (Taigh Ramey)

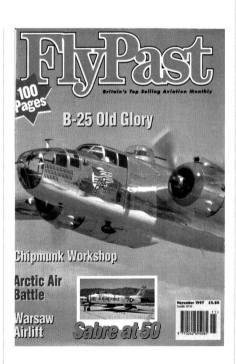

"Old Glory" in full glory on the cover of Flypast Magazine from November 1997. (Collection Wim Nijenhuis)

N7946C flew as "Dream Lover" with James Ricketts before she was purchased by the Newmans. (Collection Wim Nijenhuis)

in December 1978, to James Ricketts. With Ricketts in Stockton, California, she flew as "Dream Lover" and was damaged in a forced landing in 1987. In December 1993, she was sold to World Jet, Inc. of Fort Lauderdale, Florida. In September 1994, she was sold to Russ & Don Newman. She was delivered by road and underwent about 20,000 hours of restoration. Her first flight after restoration was in September 1995. She was named "Old Glory". In naming the airplane, Newman looked for some non-offensive nose art and something that would express what the plane was all about. "Old Glory" reflected his feelings for all military branches and for those patriots and their machines who subdued the enemies of freedom. The airplane was a star attraction at air shows across the nation since she began flying again in 1997. Newman had partnerships like that with the centre and local groups like the Confederate Air Force, Spirit of Tulsa squadron and other buffs, which makes the old airplane accessible to the public. "Old Glory" had been lent to the Tulsa Air and Space Centre and helped recruit members and raised funds for the educational programs of the museum. In 2003, the beautifully polished, natural finish B-25 was sold to John Ward Aviation Co. Inc. in Coulterville, California.

A rare picture of N9610C parked derelict at the San Fernando Drag Strip in the 1960s.

(Collection Wim Nijenhuis)

SAN FERNANDO DRAG STRIP, SAN FERNANDO, CALIFORNIA

Little is known about the B-25 with s/n 44-30585. After her storage at Davis Monthan, she got a private owner and was registered as N9610C in 1958. The purchaser from the USAF and subsequent owners are not known in the FAA files. There was a $400 lien on the airplane, which was parked at the San Fernando Airport, which was owned by the San Fernando Drag Strip and Airport Corporation. An attorney, Richard Harris, filed a deposition that proceeded with the lien sale and the airplane was transferred to the Drag Strip Corporation on 20 May, 1964. The registration was cancelled on 26 October, 1970. Probably, the airplane was scrapped after the lien sale. The San Fer-

nando Drag Strip became one of the busiest in the Los Angeles area when it opened in 1955, next to San Fernando Airport. The strip was on the northwest side of San Fernando Airport, which itself opened in 1950 on Dronfield Avenue. Fritz Burns and Bill Hannon owned the strip. They bought a spot of land that was next to the San Fernando Airport and purpose-built the drag strip. San Fernando Drag Strip ran on Sundays and closed in 1969. In the 1980s, much of the track was still there including the signage along the wash. Now, nothing remains as Home Depot occupies the end of the track, and the place has been repaved. Only the trees along the wash are still there.

Scott Aero Service Inc., Long Beach, California

A company that owned two B-25s was Scott Aero Service Inc. at Long Beach, California. These two were N7707C and N9086Z. In 1933, R.M. Scott Jr. started his own business as Scott Flying Service at Long Beach. He participated in the civilian pilot training programme. During the Second World War, he was a partner in the Visalia-Dinuba School of Aeronautics. This school conducted basic flying training for the United States Army Air Forces West Coast Training Centre (later Western Flying Training Command) under contract until it inactivated in October 1944. After the war, he went in for modifications of C-47s and general maintenance of all types of aircraft with the usual trials and tribulations of the fixed based operator. In the 1960s, Scott Aero Services Inc. was at Long Beach Municipal Airport with sales, charter, maintenance, and school activities.

N7707C

R.M. Scott purchased N7707C, a B-25J-25 s/n 44-30690, from Lampert Flying Service at Long Beach in January 1963. He owned the B-25 for only a few months and sold the ship again in May 1963 to Pacifico Imports at Long Beach. In September 1964, they sold the ship to Western Pacific Capital, Las Vegas, Nevada. One year later in September 1965, the B-25 was purchased for the second time by Scott for Scott Aero Service Inc. Finally, she was sold to Fred Lewis & Fred Zax at Portland, Oregon in February 1967. There she crashed during a forced landing and was destroyed by fire at Sitka, Alaska on 24 January, 1968.

N9086Z

A second B-25 from Scott Aero Service was N9086Z. This was a B-25J-25 with s/n 44-30613. After storage at Davis Monthan, she was sold to National Metals, Phoenix, Arizona in December 1959 and registered as N9086Z. Later, she was sold to Richard B. Johnson, Compton, California and in January 1962, she was purchased by Scott Aero Service. The company owned the plane for nearly two years. In December 1963, she was sold to Caribbean Enterprises Inc., New Orleans, Louisiana. She is reported scrapped by 1968.

The first B–25 Mitchell with a civil registration. The B–25 NX2424 of Shell in 1946 at Palisadoes Airport, Kingston, Jamica and at La Aurora Airport, Guatemala with Doolittle in front of the airplane. (USAFA)

Shell Aviation Company, New York

The Shell Aviation Company had a long tradition of promoting aviation and owned a fleet of airplanes in the 1930s. After the Second World War, they operated a B-25J. At the time, James Doolittle was the head of the aviation division. In May 1946, Lieutenant General Doolittle left the Army and returned to Shell as a vice president and director. In the 1930s, he had already been a manager of the Aviation Department of Shell Oil Company, where he conducted numerous aviation tests. Doolittle helped influence the Shell Oil Company to produce the first quantities of 100-octane aviation gasoline. High octane fuel was crucial to the high-performance airplanes that were developed in the late 1930s. He spent the 1950s as the manager of the Shell Oil Company's aviation department, vigorously promoting the production and use of 100-octane fuel. As a senior Shell executive, Doolittle flew the B-25 company airplane. Shell purchased this Army surplus B-25 in December 1946 and used her as a personal transport airplane and as flying laboratory. This was a B-25J-35 with the s/n 45-8830. At first, she was registered as NX2424, but later changed into NL2424. This was the first civil conversion of a B-25. The B-25 was used by Shell as a flying laboratory for aviation research. Purpose was to study performance in relation to fuel, engine oil and other lubricants, and hydraulic fluids. The B-25 operated at the Shell facility at the Oakland Airport in California. In 1949, the airplane reportedly went to the Babb Company at Glendale, California, and was then exported to China in 1951. Nothing further is known.

Picture of N7707C, probably taken around 1967 when she was owned by Fred Lewis & Fred Zax, Portland, Oregon.
(Collection Wim Nijenhuis)

Registered as NL2424, the airplane was used by Shell as a flying laboratory. This is the airplane with which Shell obtained the Limited Type Certificate (AL-2). The name "Shell Flying Laboratory, For Aviation Research" is written in small letters underneath the cockpit windows. (Bill Larkins)

SOUTHWEST AVIATION, LAS CRUCES, NEW MEXICO

Southwest Aviation is located in Las Cruces, New Mexico. This company primarily operates in the aircraft, self-propelled business/industry within the automotive dealers and gasoline service stations sector. Southwest Aviation is a fixed-base operator that has served the Las Cruces International Airport for many decades. It was established in 1973 and incorporated in New Mexico. The company of owner Harold Kading has provided a variety of aviation services including aircraft charter, aircraft sales, management

N3155G as "Bronco Bustin' Bomber" in September 1986 at the Topeka Air Show, Topeka, Kansas just before she was sold to Southwest Aviation.
(Robert Bourlier)

In February 1990, Southwest sold the "Bronco Bustin' Bomber" to Jerry Tepper and she flew as "Can Do". (Hemiman)

and brokerage, aircraft and jet-refuelling operations, aircraft maintenances and parts with a full-service avionics' installation. Expansion began in the early 1990s, with the addition of avionics installation and service and repair station. For some time, this company had three B-25 Mitchells with the civil registration N943, N3155G and N9936Z. Southwest Aviation is one of the two fixed-base operators (FBOs) of the airport. The other is Francis Aviation which provides aircraft fuel, charters, and facilities to handle corporate aircraft and private charters.

N943

In the 1980s, Southwest owned this B-25J-25, s/n 44-29943, with the civil registration N943. This B-25 was stored at Davis Monthan in 1958. In 1959, she was

purchased by National Metals at Phoenix, Arizona and registered as N9444Z. After she had some owners, she was derelict on fire dump at Fairbanks, Alaska in 1977. Ten years later in 1987, the hulk was airlifted by helicopter from Fairbanks and went to Southwest Aviation. She was then registered as N943. A restoration started in 1988, but she was stored and dismantled in 1989 at the Aero Trader Yard, Ocotillo Wells, California. In 2014, she went to S&R Aviation Services Inc. at Chino, California.

N3155G

After storage at Davis Monthan, the B-25J-25 with s/n 44-30832 was sold to Aviation Rental Service of St. Paul, Minnesota and registered as N3155G. She had a number of different owners and was used for mapping and survey work. She was modi-fied by installation of radar in her tail, ad-ditional seating and windows, a cargo floor, and an aerial mapping camera. Later, the additional equipment was removed and in March 1976, she was in Chino, California for a long-term restoration with Aero Trader. In June 1980, she was sold to Donald Davis of Casper, Wyoming and flew as "Bronco Bus-tin' Bomber". In September 1987, she was sold to Southwest Aviation. They ferried the airplane back to Chino in November for res-toration. In February 1990, Southwest sold the B-25 to Jerry Tepper and she flew as "Can Do". Currently, she is flown out of Penn-sylvania as "Take-off Time" by Claire Aviation Inc. at Philadelphia.

N9936Z

Like N943, this airplane was stored at Davis Monthan in 1958 and purchased by Nation-al Metals, from Phoenix in 1959. Her U.S. se-rial number was 44-30756 and she was civil registered as N9936Z. In the 1960s, she flew in Alaska with Colco Aviation Inc. at Anchor-age as tanker #3. By 1975, she was retired, stripped and derelict at Fairbanks. After changing of different owners, she finally went to Southwest Aviation in November 1985. She was stored disassembled in Alas-ka and in 1994 she was stored disassembled at the Aero Trader yard at Ocotillo Wells. In 2014, she went to S&R Aviation Services Inc. at Chino.

The overall natural aluminium fin-ished N9936Z. In the late 1970s, she was stripped and stored at Fair-banks. (Collection Wim Nijenhuis)

TELEVISION ASSOCIATES, MICHIGAN CITY, INDIANA

A B-25 used for television business was N58TA. This bomber was stored at Davis Monthan in December 1957. Circa 1958, the airplane with the USAAF s/n 44-29910 was sold to a private owner and registered as N2835G. In 1959, the B-25 was purchased by Television Associates and registered as N58TA. Television Associates of Indiana, headquartered in Michigan City, Indiana, was formed in 1947 by William Eddy. Capt. William Crawford Eddy did a bit of every-thing. He was a naval officer, television pro-ducer, inventor, engineer, and cartoonist. Eddy formed Television Associates eventu-ally with 200 employees. The firm devel-oped equipment and techniques for geo-graphical surveys using low-flying aircraft. For this, Eddy earned a private pilot license and handled the B-25 for most of the test flights. This activity grew into a four-aircraft service, mapping profiles along 25,000 route-miles all over the world, including a 3,000-mile track through Turkey, Iran, Iraq, and Pakistan. An inventive genius, Eddy led his firm in many electronic developments, including a continuous-loop, eight-track player for the Navy. In 1961, Television As-sociates was acquired by Westinghouse Air Brake Company and merged with its sub-sidiary Melpar, one of the most prominent technology firms of that time. Television As-sociates was maintained as a Melpar subsid-iary, with Eddy serving as Board Chairman and President. There were eight overseas offices, and Eddy spent much time in travel. Melpar was sold in 1972, and Television As-sociates was then dissolved.

The B-25 was sold to Lempark Inc., Miami, Florida in 1964. The next year she was dam-aged after a landing gear collapsed in Ar-gentina. She was struck-off the USCR in No-vember 1970.

N58TA of Television Associates. During TV operations in Iran, the plane was ferried via Prestwick in 1959 and in 1963 via Gatwick. (Collections Wim Nijenhuis, Tony Clarke)

Timken Roller Bearing Company, Canton, Ohio

A heavily converted and modified B-25 was used by the Timken Roller Bearing and Axle Company at Canton, Ohio. The Timken Company is a global manufacturer of bearings and related components and assemblies. In 1899, Henry Timken and his sons, Henry (H.H.) and William, established the Timken Roller Bearing and Axle Company in St. Louis, Missouri. This firm initially manufactured tapered roller bearings for use in wagons. Timken's bearings helped wagons make easier turns and also improved their manoeuvrability in other ways. In 1901, the Timkens relocated the company to Canton, Ohio, where the firm became known as the Timken Roller Bearing Company. The Timken Company's bearings became in great demand, especially as the automotive industry originated during the first decades of the twentieth century. Beginning in the 1920s, the company increasingly began to use its bearings in the manufacture of agricultural and mining equipment, and during World War II, besides providing the United States military with bearings, the business also manufactured gun barrels and steel tubing. In 1954, Timken introduced the "AP" bearing, an innovation that would have a great impact on the railroad industry. In the 1950s, Timken opened new plants in the U.S. and outside the country. Timken's sales continued to grow steadily through the first half of the 1960s. In 1970, the corporation's name was officially shortened to The Timken Company. By 1971, Timken had a total of 16 plants in operation. Today, The Timken Company has expanded across the world. The firm maintains plants in India, China, Great Britain, France, U.S.A., South Africa, and several other countries. The company topped five billion dollars in sales in 2005, and The Timken Company remains one of Canton's most profitable businesses and largest employers.

The B-25 used by Timken was s/n 41-29784, a D-model. After her USAF storage at Altus, Oklahoma, she was owned by Luis M. Borda. In August 1948, she was sold to J. Bertin Terrel of Tulsa, Oklahoma. Her first civil registration of N5078N was assigned in February 1949. She was sold later that year to the Continental Oil Company of Ponca City, Oklahoma. By May 1953, she was sold to the Timken Roller Bearing Company. At this time, her registration was changed to N122B. She was then converted into an executive transport. In September 1953, centreline wingtip fuel tanks were installed. Two months later, a complete set of new instruments, navigation and communication radios and antennas were installed. Finally, in December 1953, several other improvements and modification were made including a hydraulically operated air stair door, escape hatches, revised fuel system, modification of engine collector rings and land-

A fine example of a civil conversion. The B-25D, s/n 41-29784, with a loading door in the nose, a hydraulically operated air stair door, wing tip tanks and a full collector ring exhaust system. The airplane had eight luxurious seats and had the civil registration N122B. In the 1950s and early 1960s, she flew for many years with the Timken Roller Bearing Company of Canton, Ohio. (Collection Wim Nijenhuis)

Mid-1970s, N122B was a good-looking business plane and owned by Richard S. Dupont of Greenville, Delaware and was provided with an attractive paint scheme. From 1981, the airplane is displayed aboard the U.S.S. Yorktown at Mount Pleasant, South Carolina, as "Furtle Turtle" and registered as N2XD.
(Richard Vandervord, Janet&Frank)

ing gear improvements. These were carried out by AiResearch Aviation of Los Angeles. In December 1964, the airplane was sold to Fred Clausen of Minneapolis, Minnesota. In later years, she got other owners and her registration was changed to N2DD and to her current number N2XD. In April 1981, she was sold to the Patriots Point Naval and Maritime Foundation and was displayed aboard the U.S.S. Yorktown as "Furtle Turtle" at Mount Pleasant, South Carolina. From 2017 until 2020, the airplane was restored and displayed aboard the U.S.S. Yorktown as "The Ruptured Duck", one of the Doolittle Raiders.

Formerly called "The Furtle Turtle", the B-25 has been rechristened "The Ruptured Duck." This was part of the restoration work to convert the bomber from her former life as a private aircraft to a replica of one of the Doolittle Raid bombers.
(Patriots Point Naval & Maritime Museum)

Tripoints Associates, Miami, Florida

Tripoints, and its sister company, RWC Associates, were manufacturers' representatives dealing in reconditioned aero engines, aviation spares of all types and, during the Biafran conflict, in leasing and operating cargo carrying aircraft. Owner of the company was the American Robert W. Cobaugh. Later, Cobaugh became involved in the conflict as the driving force behind a shadowy organisation known as Phoenix Air Transport which flew arms and supplies into the beleaguered Biafra. Whether, in fact, Tripoints had any direct involvement in a B-25 deal with Biafra is not known. In 1967, the company went out of business.

N9868C

A B-25 with a curious course of life was s/n 44-29919, a B-25J-25, registered with the U.S. civil number N9868C. She was purchased at the end of 1964 by John Osterholt of Homestead, Florida and flew as "Mañana Express". In June 1967, the airplane was sold to a company called Aerographic Inc. President of this company was also John Osterholt. The airplane was operated by Tripoints Associates, Miami, Florida. In August 1967, the airplane went to Biafra. The ferry pilot of this B-25 was possibly of Cuban nationality and was accompanied by an American ex-TWA pilot and a U.S. engineer who stayed to install SNEB rockets on the aircraft. The air-

craft's history cannot be traced beyond this point until, in April 1970 when the airplane had been completely destroyed, although under what circumstances is not recorded.

N8013

A second curious B-25 operated by Tripoints was N8013. This was ex USAAF s/n 44-31491, a B-25J-30. In 1964, this former RCAF ship with the Canadian number 5245 and with a solid nose, was sold to Intercontinental Trading Corporation at Miami, Florida. The airplane was operated by Tripoints Associates. She was fitted with a long-range ferry tank and armament and delivered to Port Harcourt in October 1967. The B-25 seems to have been abandoned when it became unserviceable sometime in December 1967. When Federal forces recaptured Port Harcourt airport on 20 May, 1968, they destroyed the B-25.

This B-25J-25, s/n 44-29919, was registered as N9868C and flew in the 1960s in the U.S. as "Mañana Express". She was later used by the rebel Biafran Air Force in Africa. She is photographed here in 1967 in Miami, before being ferried to Biafra, and still in her last civil colours. (Collection Coert Munk)

A rare picture of s/n 44-31491 in Biafran colours. Before she was delivered to Biafra, she was registered as N8013 and operated by Tripoints Associates. (Collection Coert Munk)

Walter Soplata, Newbury, Ohio

Walter Soplata had a very special collection of aircraft. Soplata lived in the small town of Newbury, Ohio. He owned a farm, and through his efforts, many rare aircraft were saved from the scrap yard. Soplata, son of Czech immigrants, became a longtime union carpenter from 1947 to 1982, and was a lifelong aviation enthusiast and collector until his death in 2010. His love for airplanes began during the lean years of the Great Depression. Although a stutter would keep him from enlisting during World War II, he eventually found himself surrounded by the planes that fascinated him so much. For many years he worked in a Cleveland, Ohio scrap yard, junking thousands of warplane engines that were suddenly declared surplus. In this job, he foresaw the near extinction of the nation's historic aircraft. He felt he had to act. On his back garden in Newbury, east of Cleveland, he began his airplane collection in 1947 with a late-1920s American Eagle biplane. In the early 1950s, he started collecting the big iron. In Newbury, he had a collection of military aircraft from the 1950s, 1960s and 1970s. The collection has been kept relatively secret since Soplata's passing in 2010. The graveyard featured approximately 50 engines and 30 aircraft lying in the "plane sanctuary". Up through 1972, he collected aircraft purchased from private individuals, aviation schools, and other non-military sources. The aircraft in the Soplata collection were periodically maintained, but no major restoration work was completed. In the 1960s and 70s, Soplata used to open his property to the public, but in later years the aircraft graveyard has been kept private. Soplata had two B-25s in his collection. One was N3682G and the other was N7947C "Wild Cargo".

N3682G

The B-25-30 with s/n 44-86708, was stored at Davis-Monthan AFB in August 1958. She was declared surplus in February 1959. Initial civilian records for this B-25 are a bit sketchy due to a fire at the Federal Records Centre in Suitland, Maryland. Her civilian registration N3682G was issued in August 1959, apparently being sold to an owner in Pennsylvania. She was flown from Davis-Monthan to Pennsylvania in 1959 and it is believed that is her only civilian flight. In 1965, she was purchased by Walter Soplata of Newbury, Ohio and added to his collection. She was obtained from a scrap dealer and moved to Newbury in 25 truckloads. In August 2017, the airplane was recovered by Kevin Hooey of Callahan & Hooey insurance

and a private pilot. He acquired the plane from Soplata. Nowadays, the airplane sits in a variety of sections in Hooey's hangar in Beaver Dams and at the shop of Tom Reilly in Douglas, Georgia. The timetable is to have the airplane on wheels and airworthy in about five years.

N7947C

In January 1963, N7947C was sold to Arthur Jones and was named "Wild Cargo". But already in February, while carrying a shipment of snakes and alligators, she made a gear-up landing at Lunken Airport, Cincinnati. She was eventually auctioned off and purchased by Walter Soplata in September 1964. She was trucked to Newbury and saved from scrapping. She was stored in the open air on the graveyard near his house. In 1990, she was sold to Steven A. Detch, Alpharetta, Georgia. She was then sold to Gerald Yagen in Suffolk, Virginia in December 1997 and restored to airworthy condition. Nowadays, she is displayed in the Army Hangar of the Military Aviation Museum at Virginia Beach and still as "Wild Cargo".

In 1965, this post-war TB-25 of the USAF, s/n 44-86708, was bought by Walter Soplata. The aircraft lie rotting at the graveyard amongst overgrown foliage and scrap metal. The airplane still has the blue engine cowling rings.
(SmugMug, Collection Wim Nijenhuis))

N7947C "Wild Cargo", loaded with reptiles, made a wheels-up landing at Lunken Airport. She was eventually auctioned and purchased by Walter Soplata who took the airplane to his secret backyard warplane collection. (Roy Cochrun Collection, Bob Haney)

The airplane in May 1971. Washington Wilbert Vault Works owned the ship from April 1967 to December 1973. (Landry)

WASHINGTON WILBERT VAULT WORKS, LAUREL, MARYLAND

Washington Wilbert Vault Works is now in its third generation as a family owned business. It was founded by Royden H. Wood Sr. in 1938. He was a good friend of Wilbert W. Haase, the founder of Wilbert Vaults in Chicago. Wood received licence #13 and opened in Rockville, Maryland. Washington Wilbert Vault was one of the very first Wilbert licensees. His territory was the single largest ever granted to a Wilbert licensee. After returning from duty in World War Two, Royden H. Wood Jr. joined his father in business. The Woods family continued to work closely with Wilbert Inc. on the prod-

uct lines and in providing service, and even designed a copper-lined vault that became the precursor to what is known today as the Copper Triune. In 1957, Wood Sr. died, and his son assumed leadership of the company. Four years later, he purchased land in Laurel. The original factory was purchased from Wolfkill Vaults. A new modern plant was built, which still exists today. When Wood Jr. died in an automobile accident in 1973, his widow Lillian Peterson Wood, stepped in. She encouraged her sons Roy and Chris to learn all the jobs in the plant. The company expanded and when she died

in 1986, Roy and Chris were well trained to carry on the family business and began the third-generation tradition. So, today the company is owned by Royden H. Wood III and Christopher J. Wood. Washington Wilbert Vault Works Inc. is a provider of various burial vaults, cremation urn products and vault transfer and disinterment services to the greater Washington DC metropolitan area.

In the late 1960s, the company owned one B-25. In April 1967, Royden H. Wood Sr. bought the B-25 registered as N75755 from Jack Adams Aircraft Sales of Walls, Mississippi. The airplane was then transferred to Washington Wilbert Vault Works at Laurel. In December 1973, the company sold the B-25 to Robert W. Trainer of Lancaster, Pennsylvania. In late 1975, she was purchased by the Canadian Warplane Heritage Museum.

A picture taken at Lancaster in 1974. Robert W. Trainer of Lancaster owned the airplane a few months and in 1974, she was used for electronic flight tests when she was operated by Flight Test Engineering Co., Newark, Delaware. Note the JATO modification in the tail cone. The aircraft was modified into an executive transport. She was reported to be relatively quiet for a B-25. This was due to the extra soundproofing that had been added. An unusual modification to this aircraft was the installation of a three bottle JATO System in the rear section of the aircraft. This system remained in the aircraft for a major portion of its civilian career although the frequency of use is not known. (John Hevesi)

OTHER CIVIL B-25S

Apart from the aforementioned owners, there were several other companies and individual persons that had a B-25 at their disposal for a short or longer period. Generally, not much is known about these planes. However, some pictures of these airplanes are available and therefore, some examples of such B-25s are shown below.

N87Z in the early 1970s. At the time this photo was taken, the airplane was owned by the Global Aeronautical Museum of Phoenix, Arizona. (Dik Shepherd)

She is photographed at Kendall-Tamiami in September 1981. One year later, the natural aluminium finished bomber crashed in the Bahamas. (Rob Groenendijk)

N87Z

This was a B-25J-30 with s/n 44-86873. Glenn E. Turner and partner Kenneth Staley at Pleasant Hill, California purchased the B-25 as surplus equipment from the USAF in 1958. The airplane was outfitted to carry tropical fish from South America to the U.S. and it made several trips doing so. First becoming N9639C, she became N87Z shortly afterwards and went through a series of owners through the 1960s and 1970s. She was also used for aerial survey and test work. In August 1982, while owned by Marcelo Ortiz of Miami, she crashed during a drug run, near Long Island in the Bahamas with the loss of all three aboard.

N198W

This airplane was a J-30 model with s/n 44-30937. After storage at Davis Monthan AFB, she was sold to David White, Oxnard, California, in April 1958 and registered as N7693C. In December 1959, she was sold to R.A. Grunert, Ogden, Utah, and registered as N198W. After that, she had several owners until January 1963, when she was sold

to Borinquen Imports Co., Santurce, Puerto Rico. Probably, there she flew until 1970. Further information unknown.

N3442G

As with many other B-25s, N3442G was stored at Davis Monthan AFB in 1958. She was a B-25J-30 model with s/n 44-86715. In May 1959, she was sold surplus to United

Aerial Applicators Inc. at Papillion, Nebraska and registered as N3442G. It was planned to convert the airplane into a sprayer but at the time the B-25 spraying business collapsed, and the conversion was not completed. The B-25 sat at the airport for the next few years. In July 1968, she was sold to Midwest Seafoods Inc., Denver, Colorado supposedly for eventual conversion into a fish carrier. She was reported derelict unconverted at South Omaha Airport, Nebraska until 1975. Then she was sold to George Andres at Papillion and in January 1975, to Joseph L. Davis, Oxnard, California. However, because of some court related paperwork, the ownership was not straightened out until 1980. Davis purchased the B-25 with the intent of restoring her. Therefore, in November 1980, he made his airplane ready for a ferry flight to Aero Trader in California. In July 1986, she was sold to William R. Klaers, Apple Valley, California. From 1994, she was stored and dismantled in Aero Trader yard, Ocotillo Wells, California. She was marked as USAF BD-715. Later she was moved to Rialto for restoration. The airplane remains a candidate for a complete restoration back to fly-

A picture of N198W probably taken in the early 1960s. The buzz number BD-937on the nose is still visible. (Collection Wim Nijenhuis)

N3442G ready for the ferry flight from Omaha, Nebraska to California in November 1980. *(John Derry)*

ing condition but will remain in storage for the indefinite future.

N3443G

After storage at Davis Monthan AFB, this B-25J-25 with USAAF s/n 44-30470 was owned by Henry E. Huntington III, Carmel, California and registered as N3443G. In 1966, she was purchased by Westair Co., Westminster, Colorado. In 1969, she was purchased by Bob Gardner of the Yankee Air Club Inc. at Sunderland, Massachusetts. She was operated by Damn Yankee Air Force at Turners Falls. In August 1970, the airplane crashed while landing at Orange, Massachusetts, killing the pilot.

N3677G

This was a B-25J-30 with s/n 44-86782 and after the war converted to a TB-25N. She was stored at Davis Monthan and sold to National Metals Co., Phoenix, Arizona in July 1959. From August 1959 until 1970, she

N3443G in July 1961 and at right, owner Bob Gardner of the Yankee Air Club in front of the ship, circa 1970.

(Collection Milo Peltzer, www.warbirdregistry.org)

was owned by Skyways System Inc., Miami, Florida and registered as N3677G. In January 1977, her civil registration N3677G was cancelled.

N5857V

This was a B-25J-30 with s/n 44-30982. She was delivered to the U.S. Navy as PBJ-1J in May 1945. She was transferred to the RFC in November 1945. One year later, she was sold to Sherman Machine & Iron Works at

Picture of N3677G taken at Miami International Airport in the late 1950's or early 1960's. (Larry Johnson)

NL5857V in 1948, still in her U.S. Navy three-tone blue colour scheme. (Collection Milo Peltzer)

N9091Z at Tapachula International Airport, Tapachula, Chiapas, Mexico in 1960. (Ikeda/Pat Hughes)

N9170Z after her crash in Nicaragua in 1963. (Mark Leonard)

Oklahoma City, Oklahoma. In January 1948, she was sold to Gaylord Container Corp., St Louis, Missouri and registered as NL5857V. In December the same year, the B-25 was sold to T. J. Murrell in New York and registered as N5857V. The airplane was seized as illegal export in January 1949.

N9091Z

A B-25J-30 with s/n 44-86800. She was stored at Davis Monthan in August 1958. Then she was sold to American Compressed Steel Corp. at Cincinnati, Ohio, and registered as N9091Z. In January 1960, she was sold to ORBEC Corp. of Cleveland, Ohio. In September 1960, the plane was flown by Clyde Benton Hughes aka Pat, when she crashed near Flores, Guatemala. Hughes was involved in the Cuba situation and the circumstances of his death have never been adequately explained. He had two other partners, names unknown, that were killed with him in the crash. The large pelican tail art is because Hughes was from Louisiana. The brown pelican was adopted as the state bird of Louisiana.

N9170Z

This was a B-25J-25 with s/n 44-30187. After storage at Davis Monthan AFB, Arizona she was purchased by National Metals, Phoenix, Arizona in January 1960. She was registered as N9170Z. In February 1960, she was sold to Melvin J. Weighmann of Tampa Caribbean Enterprises, Tampa, Florida and in May 1962, she was sold to Angels Aviation at Tampa. In December 1962, she was sold to Air Trading Co., San Juan, Puerto Rico and in January 1963 to Borinquen Imports Co., San Juan, Puerto Rico. On 21 July, 1963 the airplane crashed and destroyed in Nicaragua

N9754Z in 1967 at Honolulu and the remains after the filming of "Tora, Tora, Tora" in 1969.

(Nick Williams)

During the late-1950s and into the mid-1960s, the New York Central Rail Road owned and operated #44-30996. The railroad acquired her as surplus for $3,000 from Davis-Monthan, and then re-configured her into a combination passenger and freight carrier. She can be seen here in the last days of her service in the USAF. (Collection Wim Nijenhuis)

N9754Z

This airplane was a B-25J-25 with s/n 44-30478. By 1951, she was transferred to the Royal Canadian Air Force. In 1961, she was bought by Carispac Air Transport Ltd. at Alberta, Canada and in November 1961, the B-25 was purchased by Ocean Marine Corp., Seattle, Washington and registered as N9754Z. From 1962 until 1969, she was owned by L. Frederick Pack Associates Hawaii Ltd. at Honolulu, Hawaii. Circa 1964, she was shipped to Hawaii and used in filming of "In Harm's Way". She was stolen by an intoxicated pilot and flew low over Waikiki at night. She was damaged when she struck a radio tower and landed back at Honolulu in May 1965. The airplane was parked unrepaired at Honolulu. The wrecked plane was used during filming of "Tora Tora Tora" in 1969. She was converted with a single tail to cosmetically appear as an A-20 Havoc. The airplane was in a hangar when a Zero replica was dropped on her for a scene. Her remains were eventually scrapped by 1969.

N9991Z

This B-25 was a J-30 model with s/n 44-30996. After storage at Davis Monthan AFB, she was purchased by James J. Wright, Cleveland, Ohio in October 1958 and registered as N9991Z. James J. Wright was Director of the Technical Research Department of the New York Central Railroad, located in Cleveland. The Technical Research Department was officially dedicated in May 1957. Established as an applied research labora-

Two times N9991Z.
Above: *Circa 1966 at Chicago Midway International Airport. She still has the colours of New York Central Rail Road and was painted in the same grey–and–black scheme as the locomotives of the company. She eventually became a casualty of the railroad filing for bankruptcy and was sold to private owners.*
Below: *Later, she had another colour scheme and with a solid nose. (PatB, Collection Wim Nijenhuis)*

tory, its goal was to utilize the latest technological discoveries in the modernisation of the New York Central Railroad, with the express purpose of enabling its trains to move faster, safer, and more economically. In November 1964, she was sold to Graubart Aviation, Valparaiso, Indiana. In April 1965, she was sold to Palmer Lake, Atlanta, Georgia and in February 1966, she was sold again to Graubart Aviation. Mrs. Freddy Van Dux from Chicago, Illinois bought the plane in July 1966. Finally, she was bought by Aero Service Corp., Manila, Philippines in June 1967, and she was registered as PI-C905. There she was reportedly destroyed in a storm during 1970.

MUSEUMS

As mentioned before, several civilian B-25s ended their career in various museums. Some were restored and displayed as a static airplane. Others were restored to a flying condition and were highly appreciated guests at air shows and sometimes still today. That was mainly dependent on the various objectives, locations, and financial possibilities of the museums. Many of these museums and their airplanes are described below.

AIR ZOO, KALAMAZOO, MICHIGAN

A beautiful and colourful B-25 is displayed at Air Zoo in Kalamazoo, Michigan. The Air Zoo's founding organisation, the Kalamazoo Aviation History Museum, was founded in 1977 and it was opened to the public on 18 November, 1979. However, the Air Zoo had its beginnings much earlier. Pete and Suzanne Parish shared their enthusiasm about World War II airplanes, and they had collected some airplanes. They founded the Kalamazoo Aviation History Museum in 1977. The museum began expanding its facility in 1986 and tripling its size. Completed in 1987, the expanded museum housed a larger exhibit hall, a video theatre, a new museum store and a larger library. In 1994, the museum expanded again by moving its existing Flight and Restoration Centre to a newer and larger building. The museum was then chosen as the home of the Michigan Aviation Hall of Fame. In early 1999, the name "Air Zoo" was adopted. In 2002, several additional milestones were achieved. The culmination of a yearlong process resulted in the museum becoming an affiliate of the Smithsonian Institution. In May 2004, the Air Zoo opened a new facility unlike any other in the world.

The 120,000-square-foot facility is what some call the "museum of the future." In October 2011, the Air Zoo expanded yet again by opening a 50,000-square-foot addition on its Main Campus. The expansion features new and current exhibits, aircraft, and space artefacts. The Air Zoo has more than 50 fixed and rotary wing airplanes on display, from a replica of the first Wright Flyer and World War II era fighters and bombers, to the modern spacecraft.

In 1956, N37L was converted for executive service, including an air stair and new wings with wing-tip tanks. *(Collection Wim Nijenhuis)*

At first, the Mitchell was stored outside the Kalamazoo Aviation History Museum. This picture was taken in July 1987. *(Glenn Chatfield)*

One of the airplanes of the Air Zoo is the B-25H-10 with s/n 43-4899. This airplane was delivered in May 1944. She was used as an administration airplane at the Kansas City Modification Centre. She was assigned to a replacement pool in the Philippines and consequently never saw combat.

In the 1990s, N37L was cosmetically restored and given an 8-gun solid strafer nose with the colours of the famous 345th Bomb Group "Air Apaches". Nowadays, she is displayed inside the museum.

(Dees Stribling, Air Zoo)

In November 1944, she arrived back in the U.S. and was assigned to the 1st Motion Picture Production Unit at Culver City, California for the next year. In February 1946, she was flown to the RFC at Altus, Oklahoma and followed by improperly processed sales. Finally, in March 1952, she was sold to William Rausch of Teterboro, New Jersey and by July 1955, her registration of N1582V was assigned.

The next month, she was again sold to the Le Franc Company of San Francisco, California and her registration was changed to N37L. She was converted for executive

service in 1956. The conversion included an air stair and new wings with wing-tip tanks. From 1959, she changed several times of ownership. In 1980, the Kalamazoo Aviation History Museum acquired the B-25. She was trucked to Kalamazoo. At the time, the museum just opened two years earlier, it was impractical to do much more than paint the airplane and put her outside on static display. In 1990 it was decided to upgrade the bomber from outside display to inside display with an authentic military colour scheme. She was cosmetically restored and given an 8-gun solid strafer nose with the famous "Air Apaches" colours.

AMERICAN AERONAUTICAL FOUNDATION, CAMARILLO, CALIFORNIA

A popular B-25 that millions of aviation fans have seen is a warbird named "Executive Sweet" of the American Aeronautical Foundation. This foundation is a flying museum that offers people the possibility to experience what it is like to fly aboard a real-World War II bomber.

The foundation is a group of volunteers located at Camarillo, California that have a restored and flyable B-25. The American Aeronautical Foundation was founded in 1982 to help preserve the aviation legacy of American World War II veterans and the aircraft they flew. There are pilots and maintenance personnel in this group. The AAF Group has two airplanes, a B-25 and a C-47 Dakota. Its flagship is the B-25 and has been seen at air shows, fly-ins and private avia-

tion events for more than four decades. This old warbird is a B-25J-25 with s/n 44-30801. After USAF storage she was sold to Fogle Aircraft, Tucson, Arizona in 1959 and registered as N3699G. Thereafter, she was sold

"Executive Sweet" at the air show at Miramar in 1991. She was overall aluminium finished with a beautiful nose art.

(Collection Wim Nijenhuis)

N30801 is painted in an Olive Drab/ Neutral Grey colour scheme. End 2023, the Liberty Foundation at Douglas, Georgia became her new home and once again she was provided with a new nose art.
(John Thow/Sickpixels.com)

Different nose art

In the course of time, "Executive Sweet" has had several nose art paintings. Clockwise from top left: 1974, 1976, 2004, 2012 and 1991.
(Collection Wim Nijenhuis)

a few times and flew as a sprayer. In 1968, she was sold to Filmways Inc. and flew in the movie "Catch-22". Put on sale after completion of filming in 1970, she was purchased by Ed Schnepf of Challenge Publications, Van Nuys, California in February 1972 and registered as N30801. A two-year restorative programme brought her back to a wartime J model appearance. Looking factory new in her bare metal finish, the airplane once again was armed with thirteen .50 calibre machine guns, a Norden bombsight and operable bomb bay, authentic insignia, and interior detailing down to crash axes and crew intercom. In July 1982, Schnepf donated the B-25 to the newly formed American Aeronautical Foundation Museum at Camarillo, that has been operating her ever since. The airplane has appeared in a dozen major films and numerous TV shows and commercials. In April 1992, the airplane participated in the Doolittle Raid 50th Anniversary re-enactment staged at North Island Naval Air Station in San Diego, California. This fully restored B-25 was one of the first in the warbird circuit and was an example that was later followed by many.

AMERICAN AIRPOWER MUSEUM, FARMINGDALE, NEW YORK

N2825B in the 1970s with modified solid nose and the current glass nose with nose art "Miss Hap". (Collection Wim Nijenhuis)

The American Airpower Museum in Farmingdale, New York is a museum focused on aviation situated on the former grounds of Republic Aviation. The aviation museum contains a unique collection of artefacts related to aviation, as well as a variety of aircraft that span the numerous years of history of the aircraft factory. The museum has a group of volunteers which includes both former Republic workers and veterans of all branches of the military. Exhibits and aircraft range from WWI to the Iraq War.

Among others, the museum operates an authentic Republic P-47D fighter as part of its fleet of aircraft. Visitors can also enjoy a flight experience aboard a C-47 aircraft, one that flew on D-Day during the invasion of Normandy. The historic aircrafts taxi to the same hangars and runways that once sent Thunderbolts off to battle. These aircrafts recreate those troubling years of war and provide visitors with a rare opportunity to view the airplanes flying through the sky as they did in the past.

One of the main strategic goals of the American Airpower Museum is preserving the legacy of those brave men and women who made the ultimate sacrifice in defending the liberties of Americans. The other important mission of the museum is the education of newer generations regarding the heroism, valour, and courage that is shown in the citizen soldiers of the United States through the presentation of historic operational aircrafts, as well as related programs and exhibits.

The B-25 with s/n 40-2168 of the American Airpower Museum is the oldest known B-25 in existence. She was one of the very early production examples having straight wings with a constant dihedral all the way to the wing tips. During World War II, the airplane was rebuilt and modified for use as a staff transport for General Henry "Hap" Arnold. After the war, she served as an executive transport for various companies. In 1983, her modified solid nose was replaced with a standard greenhouse nose. This ship is registered as N2825B and since 2002 operated by the museum and flown as "Miss Hap".

BIG KAHUNA'S WATER AND ADVENTURE PARK, DESTIN, FLORIDA

Not a museum, however, the Big Kahuna's Water and Adventure Park at Destin, Florida displayed a real B-25 like a museum. In the park a B-25 Mitchell bomber was placed on a pole. This airplane was s/n 44-30947, a B-25J-30 (in literature many years incorrectly referred to as 44-86844). Big Kahuna's Water and Adventure Park is located in Destin, Florida and opened in 1986. It is the largest water park on the Emerald Coast. Among its tropical landscaping and cascading waterfalls there are more than 40 water attractions including four children's areas, high speed body flumes and thunderous white water tube rides which empty into meandering lazy rivers. The adventure park is complete with one of the largest regional tropical mini-golf courses and two adrena-

"Miss Hap" during the B-25 Gathering at Grimes Field in April 2017. (Wim Nijenhuis)

In 1998, N92880 was displayed on a pole above the water rides at the park.
Right: *Photographed after her right vertical stabiliser and rudder were blown off in Hurricane Ivan in 2004.* (Collection Wim Nijenhuis, Paul Robbins)

CAVANAUGH FLIGHT MUSEUM, ADDISON, TEXAS

line pumping thrill rides. The B-25 bomber was delivered to the USAAF with s/n 44-30947. In October 1951, she was taken on strength by RCAF Air Defence Command. After some modifications and service, she was stored at RCAF Station Lincoln Park, Alberta in December 1960. She was registered as CF-NTP for her ferry from Lincoln Park to Calgary, Alberta, in October 1961. In 1962, she was sold to Columbus L. Woods/Woods Body Shop, Lewistown, Montana, and registered as N92880. In April 1963, the airplane was purchased by Arthur Jones of Slidell, Louisiana. She was noted derelict at the New Orleans-Lakefront Airport, Louisiana in 1977. She was sold to M.W. Hamilton, Petal, Mississippi and then to Jay Wisler at Tampa, Florida in September 1997. She was sold to Big Kahuna's Waterpark in 1998. There she was displayed on a pole above the water rides in an overall yellow colour and numbered 362436. On her nose was painted a pin-up and the name "Jungle Queen". In 2004, she was damaged and lost her right vertical stabiliser and rudder in Hurricane Ivan. In 2005, she was removed from the pole and sold to Glen Hyde, Roanoke, Texas. She was transported by truck to Roanoke and stored pending restoration. Nowadays, her forward fuselage section is being used to rebuild the B-25 "Sandbar Mitchell", s/n 44-30733, at the Warbirds of Glory Museum at Brighton, Michigan.

The Cavanaugh Flight Museum at Addison, Texas is mostly a personal collection of aviation super-fan Jim Cavanaugh. Tucked away in the hangars at Addison Airport, most of the vintage planes are fully airworthy. Growing up with an aviation father, Jim Cavanaugh was predisposed to become involved in airplanes himself. In 1968, he was working nights as a hotel accountant. Realising that janitorial service was a sector that would be required regardless of economic upswings or downswings, he began making contracts for janitorial service by day. Within a year, Jim Cavanaugh had established Jani-King and, thus, entered commercial cleaning services in a big way. Jani-King International, Inc. is now an American chain of cleaning service franchise. The company provides cleaning for public spaces such as offices, retail spaces, hotels, and stadiums. The success of the company allowed him to finally pursue his love of flying. He used a corporate jet for business travel, but he has always had an eye for warbirds. He

The original "Barbie III" of the 1st ACG in flight during WWII. In 1988, during restoration, bomber N5548N was shown at an air show at Rockford, Illinois.
(USAF, Glenn Chatfield)

N5548N

In the 1950s and 1960s, the B-25H-1 s/n 43-4106 with civil registration N5548N was used by Bendix Aviation for many years as a test aircraft. After Bendix Aviation sold the B-25 off, she went through several owners. In 1969, she eventually was purchased by D. Richard Lambert of Plainfield, Illinois. In August 1981, the bomber was sold to Weary Warriors Squadron of Chicago, Illinois. After two engine failures, they were able to finally fly the bomber off Lambert's farm in 1981. From there, the group embarked on a full restoration of the aircraft. Finally, in 1992, after a restoration of ten years, the bomber was restored with the gun nose with 75mm cannon. They had purchased a group of surplus B-25 parts for use with their aircraft. Among those parts was the majority of a B-25H nose section. The remainder of the nose was salvaged elsewhere, restored, and fitted to the airframe. After a 10-year restoration, the Mitchell operated as "Barbie III" and did her first air show in March 1994. Although the bomber did not serve in combat, the restoration group decided to paint the airplane in the colours of a combat veteran B-25H.

A magnificent aircraft, the restored warbird "Barbie III" in 2017. (Dennis Nijenhuis)

In April 2017. This is the only flying B–25H in the world. (Wim Nijenhuis)

purchased a P-51 Mustang and placed the Mustang in a hangar along with some other lightplanes. Cavanaugh's collection of airplanes grew with the addition of a T-6 and an F-86. What started out as a private hobby soon grew into a serious collection. The Cavanaugh Flight Museum opened in October 1993. The collection includes aircraft from World War I, World War II, the Korean War, and the Vietnam War. The WWII collection includes some very rare combat veteran airplanes. The museum has an extensive maintenance facility. Since it opened its doors, the museum has expanded its facilities, but it remains at the same place. The Cavanaugh collection includes two airworthy B-25 Mitchells and are a B-25H and a B-25J.

Picture of #44–28925 taken in November 1972 when she was being used as a war memorial at the Forest Lawn Garden Cemetery near Pittsburgh, Pennsylvania. Later, she was displayed on a pole as "Daisy Jean". In 1984, she was removed and trucked to Florida for rebuild. (Bill Larkins via Nick Veronico)

The artwork of Ray Kowalik. "How Boot That?!" during World War Two and from the restored airplane at the Cavanaugh Flight Museum. (Collection Wim Nijenhuis)

Their choice was the B-25H "Barbie III" as flown by Lt. Col. Robert T. Smith, Commanding Officer of the bomber section of the 1st Air Commando Group. His airplane was named in honour of Smith's wife, Barbara Bradford, whom he married shortly before Smith departed for the CBI in June 1943. In June 2009, History Flight Inc. purchased the B-25 and in 2015, she was purchased by Cavanaugh Flight Museum and currently this magnificent airplane is the world's last airworthy B-25H.

N7687C

The B-25J with s/n 44-28925 of Cavanaugh is N7687C. This airplane is the crown jewel of the museum's collection. She was the most original, flying B-25 anywhere in the world. In 1944, she was assigned to the 380th Bomb Squadron of the 310th Bomb Group. The airplane arrived in Italy shortly after her completion. From the fall of 1944 through late spring 1945, she completed more than eighty combat missions over northern Italy, southern Austria and what then was Yugoslavia. After the war, she returned to the U.S. and continued to serve with the USAF. The USAF struck off the bomber from its inventory in 1958. She was sold to Parsons Airpark at Carpenteria, California and registered as N7687C. In 1968, the bomber appeared in the film "Catch-22". In May 1971, she was sold to Mr. John Neel of Washington, Pennsylvania to be displayed at Forest Lawn Gardens in McMurray, Pennsylvania, a Veterans Cemetery. In 1984, she was sold to Harry Doan, Daytona Beach, Florida and was stored disassembled at Kissimmee, Florida. In November 1992, she was sold to the Cavanaugh Flight Museum. The museum restored the B-25 with the help of Aero Trader in 1995. The restoration of the bomber was complete in every detail and all the plane's systems are fully operational. Ray Kowalik, the same artist who first created the plane's distinctive nose art of "How Boot That!?" in December 1944, faithfully reproduced it as part of the restoration effort. Although maintained airworthy, "How Boot That!?" is currently considered "too valuable to fly".

"How Boot That!?" in 2014. The restoration of the bomber was complete in every detail and all the plane's systems are fully operational.

(David Stubbington)

CHAMPAIGN AVIATION MUSEUM, URBANA, OHIO

Located at Grimes Field in Urbana, Ohio, the museum's purpose is to restore and preserve historical airplanes specifically but not limited to the airplanes of World War II. Organised in 2005, the Champaign Aviation Museum is a flying museum and is working on restoring a B-17 and collects, exhibits, and preserves aircraft of all types for viewing by the general public. The museum opened in 2007, and a group of volunteers has been working on restoration of planes since before that, in 2006. The museum's biggest project is rebuilding B-17 "Champaign Lady" from parts of several other B-17s, with much of it scratch-built by volunteers. Its goal is to make the B-17 bomber flyable. The reconstruction had its genesis with the Shiffer family, owners of Tech II Inc., an Urbana plastics manufacturer. They bought the B-17 from Tom Reilly's Vintage Aircraft in Kissimmee, Florida. Reilly's airplane had been damaged in a crash landing several years earlier and was in far from flyable condition. The Shiffers brought the damaged bomber to Urbana in 2005. They set up a non-profit organisation and place to restore the plane and formed thereby the Champaign Aviation Museum. Apart from the B-17, the museum has five other airplanes including a B-25 Mitchell.

Grimes Gathering of B-25s. In 2012 and 2017, Grimes Field welcomed B-25s that participated in ceremonies commemorating the 70th and 75th Anniversary of the Doolittle Raid on Japan. This is the cover of the Commemorative Booklet of 2012.

(Collection Wim Nijenhuis)

N744GC in her dark green colour scheme and named "Carolina Girl" before she was bought for the Champaign Aviation Museum. (Michael D. Brooks)

This Mitchell is a B-25J-15 with s/n 44-28866, accepted in July 1944. She was used for advanced twin-engine pilot training, administrative duties, utility transport and transfer of U.S. Air Force personnel to base assignments. In 1957, she was stored at Davis Monthan. In January 1958, the bomber was purchased by Boeing and civil registered as N5277V. In the 1960s she was sold to Canadian companies and registered as CF-OND. She flew as a tanker and by October 1993, she returned to the U.S. and was sold to BHA Leasing of Carmel, Main. Her registration was changed to N225AJ. From 1999 to 2004, she was owned by Peddycord Foundation for Aircraft Restoration, Asheboro, North Carolina. She was restored to airworthy condition and named "Carolina Girl". Her first flight was in October 2004 and she was registered as N744CG. In November 2004, she was sold to Branson's Bombers Ltd., Greensboro, North Carolina and flew as "Hobbes". By 2008, the bomber was purchased by Jerry Shiffer at Urbana. She was loaned to the Champaign Aviation Museum and currently flies in a dark green camouflage scheme as "Champaign Gal".

In 2012, the bomber performed this flyby over the National Museum of the U.S. Air Force in honour of the 70th reunion of the Doolittle Tokyo Raiders. (top) Five years later, "Champaign Gal" in April 2017, at the 75th Doolittle Tokyo Raiders Anniversary at Dayton, Ohio. (USAF/Tech. Sgt. Bennie J. Davis III, Wim Nijenhuis)

COLLINGS FOUNDATION, STOW, MASSACHUSETTS

The Collings Foundation is a private non-profit educational foundation located in Stow, Massachusetts. It is founded in 1979 by Robert F. Collings and Caroline Collings. Bob Collings is an entrepreneur and passion-

ate automotive and aviation buff. Collings made his fortune in the early 1970s with Data Terminal Systems, which manufactured the first stand-alone, electronic cash registers. Later, he co-founded Resource Dynamics Inc., a proprietary facilities asset planning and management system. His entrepreneurial and financial success led him to acquire dozens of extraordinary automobiles. One thing led to another, and before long, he had the first hints of a proper museum collection on his hands. The purpose of the Collings Foundation is to organise and support "living history" events and the presentation of historical artefacts and content that enable Americans to learn more about their heritage through direct participation.

The Collings Foundation is headquartered from a small private airfield in Stow, Massachusetts that includes a small museum that opens for special events and pre-scheduled tour groups. The organisation also has a satellite operation base in Houston, Texas located at Ellington Field primarily housing the Korean War and Vietnam War jet aircraft and helicopter collection. Since 1989, a major focus of the Foundation has been the "Wings of Freedom Tour" of WWII aircraft. This tour showcases three fully restored bomber aircraft: a B-24 Liberator, a B-17 Flying Fortress and a B-25 Mitchell along with a dual-control TF-51 Mustang fighter. But unfortunately, the beautiful restored B-17 crashed on 2 October, 2019 at Bradley International Airport in Windsor Locks, Connecticut, while attempting to return shortly after take-off. The aircraft was destroyed in the crash, and seven of the thirteen people on board were killed. The "Wings of Freedom Tour" has two goals: to honour the sacrifices made by U.S. veterans that allow the Americans to enjoy their freedom; and to educate the visitors, especially younger Americans, about their national history and heritage. The Foundation encourages peo-

N3476G in the former Desert Sand scheme as "Hoosier Honey".

(Larry Johnson)

ple to tour the airplanes, talk to the veterans who come to visit the aircraft, and participate in a "flight experience". The American Heritage Museum in Stow includes an aviation museum and a vintage automobile and race car collection of more than 75 vehicles and tanks. The aviation museum is home to several the Foundation's smaller aircraft. During the past 30 years, the Foundation has undertaken and completed many restoration projects.

The B-25 operated by the Collings Foundation is N3476G. This former tanker was purchased by the Collings Foundation in 1984 and over the following two years, the aircraft was restored by Tom Reilly Vintage Aircraft and was initially displayed in a Desert Sand scheme as the "Hoosier Honey", a composite B-25 representing aircraft 7A with the 12th Air Force in service in North Africa and Italy in 1944. The "Hoosier Honey" was operated by the Collings Foundation in the Boston area over the next ten years flying to air shows and events staffed by a volunteer crew of enthusiasts and veterans. In 1994 the B-25 was flown to Houston, Texas to join the volunteer group and aircraft

N3476G "Tondelayo" of the Collings Foundation with beautiful nose art and the mustang of the 500th BS, 345th BG on her vertical tails.

(Collection Wim Nijenhuis)

composing Collings Foundation West based at Ellington Field. In a move to preserve the aircraft and refresh it from its years of operation, the Collings Foundation took the airplane to Chino, California for overhaul by B-25 restoration expert Carl Scholl of Aero Trader, Inc. in late 2001. After work was completed, the B-25 was flown to Midland, Texas to be repainted as "Tondelayo". This airplane flew in the 345th BG "Air Apaches", 500th BS of the 5th Air Force in the Pacific Theatre against targets in New Guinea. Currently, N3476G is one of the three bombers of the "Wings of Freedom Tour".

COMMEMORATIVE AIR FORCE, DALLAS, TEXAS

The cover of a 1994 promotional flyer of the CAF with its B-25 "Show Me" and the world's only flying B-29 "Fifi" at the time.
(Collection Wim Nijenhuis)

"Yellow Rose" in her early days with the CAF Alamo Wing and still registered as N9077Z. In 1978, she was called "Yellow Rose" after Emily Rose from San Antonio.
(Richard Vandervord)

Not a museum with static aircraft. No, the Commemorative Air Force is mainly a "flying" museum. The Commemorative Air Force was founded to acquire, restore and preserve in flying condition a complete collection of combat aircraft which were flown by all military services of the United States, and selected aircraft of other nations, for the education and enjoyment of present and future generations of Americans. In 1957, Lloyd Nolen and four friends from the Rio Grande Valley in Texas, purchased a P-51 Mustang. They formed a loosely defined organisation to share the pleasure and expense of maintaining this warbird, known as "Old Red Nose".

In 1958, the group made its second purchase of two Grumman F8F Bearcats. At this point, the mission of the CAF became clear: save an example of every aircraft that flew during World War II. By 1960, the group began to search seriously for other World War II aircraft, but it quickly became apparent that few remained in flying condition and almost all the warbirds were gone. In September 1961, the CAF was chartered as a non-profit Texas corporation to restore and preserve World War II-era combat aircraft.

By 1963, they had completed a collection of 10 fighters, when they purchased a B-25 Mitchell bomber and found themselves in the bomber business as well. Their first air show was held on 10 March, 1963. In 1965, the first museum building was completed at old Rebel Field, Mercedes, Texas. The CAF created a new Rebel Field at Harlingen, Texas, when they moved there in 1968, occupying three large buildings. The CAF fleet continued to grow and included heavy bombers such as the B-17, B-24 and B-29. They also had medium bombers like the A-20 and B-25. The CAF's fleet of historic aircraft was known as the CAF Ghost Squadron. The year 1991 marked the beginning of a new era for the CAF with the opening of the new Midland headquarters and museum with better facilities to preserve the CAF fleet which then numbered

"Yellow Rose" of the Central Texas Wing at the Airsho at Midland, Texas in October 1994.
(Wim Nijenhuis)

fly more than a few hours from their home base, Airsho is also an opportunity for CAF members to meet up. The CAF Airsho is one of the largest warbird air shows in the world.

Over the years, the CAF has owned B-25 Mitchells with the following civil registration numbers: N25YR (N9077Z), N125AZ (N9552Z), N345TH (N3481G), N3676G, N5865V, N9643C, N27493 (N3160G).

N25YR

After her USAF service, this airplane, a B-25J-5 s/n 43-27868, was stored at Davis-Monthan. She was released by the military in December 1959 and sold to Fogle Aircraft Company of Tucson, Arizona. She was again sold in February 1960 to Dothan Aviation Corp. of Dothan, Alabama. The next month, she was registered as N9077Z. In January 1962, she was modified for agricultural spraying and dusting and in October 1975, she was sold to John Stokes of San Marcos, Texas. In September 1977, she was sold again to Charles Skipper, Charles Becker and Jack Jones at San Antonio, Texas. Finally, she was bought by the Confederate Air Force in July 1979 and her registration was changed to the current N25YR. She is based at the Central Texas Wing at San Marcos, Texas and still flying as "Yellow Rose".

N125AZ

The B-25J-10, s/n 43-35972, is now flying as "Maid in the Shade". She was delivered in June 1944 and did service with the 319th Bomb Group at Corsica. She returned to the U.S. in July 1945 and returned to service duty with administrative base units. Eventu-

135 aircraft. The organisation was originally known as the Confederate Air Force. Following a membership vote in 2001 and made effective on 1 January, 2002, the organisation is now called the Commemorative Air Force. Collecting aircraft for more than half a century, the CAF now ranks as one of the largest air forces in the world. Today, the Commemorative Air Force operates a fleet of more than 175 vintage aircraft that are used to inspire and educate new generations of Americans on the value of military aviation in assuring the nation's freedom. The CAF's fleet is maintained and flown by the organisation's volunteers. The CAF has approximately 13,000 members.

The Headquarters of the CAF is located in Dallas, Texas. The CAF flies its aircraft during air shows and other public events, hosts various community institutions, provides guidance and policy for the safe operation of its aircraft. The organisation also provides information and advice to individuals and units to assist them with their aircraft operations. The CAF has many wings and squadrons. Starting in 2013, a limited number of larger units may be designated as an

"airbase." The first is Airbase Arizona, located at Falcon Field in Mesa, Arizona and redesignated in June 2013. Most CAF units are in the United States, but there are four outside the country. A yearly event is the Airsho at Midland International Airport, showcasing the CAF's aircraft. Because its aircraft tend to be spread out over large geographic distances, and most squadron aircraft rarely

The ex-insecticide spraying N9552Z had just been ferried to the Confederate Air Force when photographed in November 1981. On the nose is hand-painted "I see a problem". The name was prophetic because the airplane did not complete her rebuild at Mesa for over 25 years, flying again in May 2009. (Geoff Goodall)

In October 1981, the N9552Z was donated to the Confederate Air Force at Harlingen, Midland, Texas. In December 1999, the airplane went to the CAF Arizona Wing at Mesa and registered as N125AZ. She was restored to airworthy condition and named "Maid in the Shade". Here she is seen at the 2011 air show at Naval Air Facility El Centro, California.
(Ingo Warnecke via airport-data.com)

ally, she was retired to storage in April 1958. In January 1960, she was sold to National Metals of Phoenix, Arizona and registered as N9552Z. In the same year, she was sold to Dothan Aviation in Dothan, Alabama and converted with tank and spray bars. In 1975, she was sold to John Stokes of Cen-Tex Aviation from San Marcos, Texas. She was sold again in August 1979 to Henry W. Fisher, Donald W. Ericson, and Robert E. Thompson at Blooming, Michigan. In October 1981, the B-25 was donated to the Confederate Air Force at Harlingen, Midland, Texas. In December 1999, she went to the CAF Arizona Wing at Mesa, Arizona and was registered as N125AZ. She was restored to airworthy condition and named "Maid in the Shade".

N345TH

The B-25J-30 s/n 44-31385 is nowadays flying as "Show Me". The bomber was deliv-

ered in June 1945. She was used as a trainer until October 1958 when she was put into storage at Davis Monthan. In May 1959, the airplane was sold to Ray Karrels from Fort

Washington, Wisconsin, who had a local insurance agency. Her civil registration was assigned as N3481G. The intended use, a corporate aircraft for his insurance firm, never materialised and in October 1963, she was sold to Northwest Development Company at Kohler, Wisconsin. In November 1965, she was transferred to Community Credit Cor-

N3481G in her first civil years. She still has the USAF buzz number BD-385 visible on the rear fuselage. (Collection Wim Nijenhuis)

Top: *Before she found her home with the CAF Missouri Wing, she was owned by Jack Rhoades Inc. of Seymour, Illinois.*
(www.warbirdregistry.org)

Nowadays, the airplane is registered as N345TH and named "Show Me". Here she can be seen in flight in 2016 and at Grimes Field, Urbana, Ohio in April 2017.
(Wim Nijenhuis, CAF Missouri Wing)

Arizona, she was fitted with a 1,170 gallon tank in the bomb bay. In June 1967, she was acquired by the CAF at Mercedes (later Harlingen). In the CAF she flew as "Rowdy's Raider". At the end of the 1970s, she was deemed unsuitable for restoration to airworthy condition and major components were not suitable for spares. All usable instruments and controls were removed by the CAF and she had been on static display at Harlingen for many years. In October

poration and they sold her to John Lowe of Riverside, Illinois. She was then purchased by Jack Rhoades of Seymour, Illinois in January of 1966. In 1969, the aircraft was damaged by a windstorm while on the ground and was stored at Seymour. Fortunately, she was restored and flew again by 1976. She found her home with the Missouri Wing of the Confederate Air Force in 1982 and was officially renamed "Show Me" with a civil registration of N345TH.

N3676G

N3676G served as a fire tanker before being acquired by the CAF in the 1960s. This B-25 Mitchell flew for several years with the CAF at Harlingen. This was a B-25J-20 with s/n

"Rowdy's Raider" in flight and in 1980, in partially demolished condition as static display at Harlingen. Note the early CAF markings.
(Collection Wim Nijenhuis)

44-29835 and was stored at Davis Monthan in 1958. In July 1959, she was sold to Donald Aircraft at Tucson, Arizona and registered as N3676G. Afterwards, she changed a few times of ownership and in July 1961, when owned by Howard K. Roth from Sedona,

1980, the airplane was trucked to Lackland AFB, Texas and dismantled for restoration. She was dedicated at Lackland in December 1980 and today is a static display at the Airman Heritage Museum in Lackland.

N3676G in her final stage as static display at Lackland AFB, Texas, photographed in October 1994. (Wim Nijenhuis).

only known flying PBJ-1J. The B-25 is painted in the colours of a Marine Corps PBJ-1J, named "Semper Fi" with Globe and Anchor painted on the port side of the nose.

N9643C

The B-25J-30 with s/n 44-86758 is now flying as "Devil Dog". She was delivered in June 1945 and placed into storage. She was removed from storage in June of 1951 and modified for maintenance and converted to a TB-25K. She was used at the radar intercept training facility. In December 1957, she was flown back to storage at Davis Monthan. In April 1958, she was sold to Arrow Sales of North Hollywood, California. Her civil registration was assigned as N9643C. In August 1958, she was sold to H. H. Coffield of Rockdale, Texas. He removed the radar trainer K nose and replaced it with a solid nose. Between 1961 and 1965, she was used to transport cargo. She was again sold in February 1976 to W.A. Wooten and V.E. Thorpe of Harlingen, Texas. Two years later in February 1978, she was sold to Three Point Aviation, Belton, Missouri and in November 1980, sold to the CAF, Harlingen. Between 1976 and 1980, she was restored at Rockdale, Texas. Her final restoration was done at Harlingen and the first flight after restoration was at Harlingen in July 1981.

N5865V

This B-25J-30 s/n 44-30988, was registered as N5865V. In the 1970s, this airplane was restored by Tom Reilly at Kissimmee, Florida and flew as "Big Ole Brew 'n Little Ole You". In 1988, she was sold to Craig Tims of Roanoke,

ing. The fuselage interior was stripped, and modifications from the airplane's corporate career were removed. Many of the airplane's corroded internal formers, stringers and longerons had to be replaced, and often they were of unusual curves and shapes that required expensive subcontracting. On 15 May, 2016, after a 23 year restoration, she became the

Left: Ready for her first flight on 15 May, 2016 after 23 years of restoration. (Robert Shellabarger)

"Semper Fi" at Grimes Field, Urbana in April 2017. She is painted in the colours of a Marine Corps PBJ-1J and with Globe and Anchor painted on the port side of the nose. (Wim Nijenhuis)

Texas. She was transferred to the CAF at Harlingen in June 1988. In September 1991, she was transferred to the American Airpower Heritage Flying Museum at Midland, Texas and in 1993 assigned to the Southern California Wing of the CAF, Camarillo, California. She flew to Camarillo from Midland in April 1993. The airplane was named "Pride of the Yanks". By 1994, a long restoration was started by the Southern California Wing. Though some disassembly and clean-up work were done, serious restoration began only in 2003 that included removal of the engines, the wings outboard of the engines, and all fuel, oil and hydraulic tanks and plumb-

Wheels up, "Devil Dog" during a take-off in 2016. (Collection Wim Nijenhuis)

In August 1991, she was transferred to the American Airpower Heritage Museum. In November 2001, she was damaged during a landing accident at Lancaster, Texas. Starboard wing, propeller and engine sustained damage that was repaired. By 2009, she was grounded in need of restoration. Major work was completed. During 2016 winter maintenance, she received a new coat of paint. She is flown by the CAF Devil Dog Squadron at Georgetown, Texas as "Devil Dog" in a dark blue USMC PB-1J paint scheme. The "Devil Dog" is painted to represent ship number three of the VMB-612.

The old nose art photographed in 1985 and the art as nowadays applied to her nose.
(Collection Kevin Trotman, Collection Wim Nijenhuis)

N27493

The B-25J-20 with s/n 44-29869 is now flying as "Miss Mitchell". She was delivered to the USAAF in November 1944. After the war, she was used for training radar intercept crews in airborne radar systems. By December 1957, she was flown to storage. In Oc-

tober 1958, she was sold to Lysdale Flying Service in St. Paul, Minnesota and she was registered as N3160G. From 1963 until 1970, she was stored. In June 1970, she was sold to Robert Kundel of Rice Lake, Wisconsin. By 1974, she was located in Anoka City, Wisconsin with an Olive Drab colour scheme with yellow cowlings. In December 1978, she was sold to the CAF in Harlingen, Texas. She was assigned to the Minnesota Wing for restoration and brought to Flemming Field in South St. Paul, Minnesota in September 1978. She was restored to honour a B-25J that served with the 310th Bomb Group, 380th Bomb Squadron during 1944-1945 in the Mediterranean Theatre of Operations. During her restoration, her current nose art was painted by Ray Kowalik, who painted the same nose art on the original "Miss Mitchell". In May 1991, her registration was changed to the current number N27493. Her first flight after restoration was on 18 April, 1992 on the 50th Anniversary of the Doolittle Raid.

Picture of the ship taken in the late 1970s. The B-25 "Rowdy's Raider" is visible on the left.
(Collection Wim Nijenhuis)

N27493 "Miss Mitchell" in October 1994 at Midland, Texas and in April 2017 at Grimes Field, Urbana. *(Wim Nijenhuis, Dennis Nijenhuis)*

The original "Miss Mitchell" of the 310th BG in the Mediterranean Theatre of Operations and the ship of the CAF Minnesota Wing. Almost 50 years difference, but both nose art paintings are made by the same man, Ray Kowalik.
(www.warwingsart.com/12thAirForce/planes7.html via Jeff and Rick Wolford, Collection Wim Nijenhuis)

DELAWARE AVIATION MUSEUM FOUNDATION, GEORGETOWN, DELAWARE

The Delaware Aviation Museum Foundation is dedicated to promoting and preserving the history of aviation, focusing on aviation in Delaware and the Delaware Coastal Airport. The museum's 10,000 square foot hangar houses a display of classic, vintage, and military aircraft and aviation artefacts as well as an extensive reference library. The foundation was founded in 2004 at the Delaware Coastal Airport in Georgetown, Delaware. The museum has a number of aircraft in its collection, ranging from training aircraft to attack aircraft to the centrepiece of the collection, the B-25 "Panchito". Director of all this activity is Larry Kelley, who ended up here after three decades

In the late 1940s and early 1950s, #44-30734 was operational with the 3575th Pilot Training Wing at Vance AFB, Oklahoma. She had buzz number BD-734. **Right:** N9079Z photographed outside the SST Aviation Museum in 1977. *(Collection Wim Nijenhuis, John Hevesi)*

in the pharmaceutical industry. Although Kelley himself is not a military veteran, his fascination with World War II aviation history goes back to his childhood. His father was a World War II veteran, having served in the jungles of New Guinea and the Philippines. The museum uses funds from flight training in the B-25 to help pay expenses for maintaining it. Orientation flights, type ratings, and second-in-command type ratings are among the options for those looking to experience the heavy warbird. The museum campaigns the bomber to dozens of events each flying season, helping to educate the public about World War II and the role that aviation played in shaping the future.

The museum's B-25 is N9079Z, a J-25 model with s/n 44-30734. This B-25 was delivered to the USAAF in February 1945. After the war, she was used by the USAF as a trainer and she flew with the Air National Guard. In May 1958, she was stored at Davis-Monthan, Arizona. In December 1959, the airplane was struck off from the inventory as surplus and her civilian life began when she was sold to Donaire Inc. at Phoenix, Arizona. She was registered as N9079Z and flew as a tanker. In the 1960s, she had different owners and received the name "Big Bertha". By 1974, the bomber was getting very weary and corrosion from the chemicals was taking its toll and she was donated to the SST Aviation Museum in Kissimmee, Florida. After the museum closed, she was sold to Pat O'Neil, Robert Bolin and Jack Myer of Wichi-

ta Falls, Texas in October 1983. The bomber was restored by Tom Reilly of Kissimmee and received the nose art and markings as "Panchito" from the 396th Bomb Squadron, 41st Bomb Group. In the early 1990's, Rick Korf bought "Panchito" and operated her from the National Warplane Museum in Geneseo, New York. He moved "Panchito" to the Valiant Air Command in Titusville, Florida in the late 1990s. In 1997, she was purchased by Larry Kelley, owner of Rag Wings & Radials Aircraft Leasing LLC, Wilmington, Delaware. She now resides at the Delaware Aviation Museum in Georgetown.

EAA Aviation Museum, Oshkosh, Wisconsin

The EAA Aviation Museum has a large collection of historic aircraft. One of these is an extraordinarily rare B-25H. The EAA Aviation Museum is a museum dedicated to the preservation and display of historic and experimental aircraft as well as antiques, classics, and warbirds. The museum is located in Oshkosh, Wisconsin, home of the museum's sponsoring organisation, the Experimental Aircraft Association (EAA), and the organisation's EAA AirVenture Oshkosh that takes place in late July/early August. The EAA AirVenture Oshkosh is the world's biggest fly-in and air show. Formerly, the museum

was called the EAA AirVenture Museum. The museum has over 200 aircraft, indoors and outdoors, and other exhibits. The EAA is a growing and diverse organisation of members with a wide range of aviation interests and backgrounds. Founded in 1953 by a group of individuals in Milwaukee, Wisconsin, who were interested in building their own airplanes, EAA expanded its mission of growing participation in aviation to include antiques, classics, warbirds, aerobatic aircraft, ultralights, helicopters, and contemporary manufactured aircraft.

Two times "Panchito". At left, the original one of Capt. Don Seiler of the 41st BG, 396th BS, in 1944. Capt. Seiler named his plane "Panchito" after the feisty Mexican rooster from the 1943 animated musical "The Three Caballeros". At right, as warbird about 75 years later with the same nose art. (via Larry Kelley, Wim Nijenhuis)

The B-25 of the EAA Aviation Museum rolled off the North American factory production line at Inglewood, California in 1943 and flew with the U.S. Army Air Forces with s/n 43-4432. In June 1947, she was registered as NL90399, later N90399, and in February 1952, she was registered as N10V. She passed through a series of owners including Husky Oil Company, Barbara Hutton, and Tallmantz Aviation. She flew in the movie "Catch-22" as "Berlin Express". In May 1971, she was sold to Dr. William S. Cooper of Merced, California who donated her to the EAA Air Museum Foundation one year later. A full restoration was started in 1975 and she was flown as "City of Burlington" with the standard B-25J greenhouse nose. During a maintenance flight in the 1980s, she suffered a main gear failure on landing. Some cosmetic restoration was done, and she was put on display. In January 2015, she was taken off display for a complete restoration. The restored "Berlin Express" engines were started in March 2019 for the first time. On 20 April, 2019, the B-25 made its first flight in decades after a thorough restoration that took nearly four and a half years. When the restoration is complete, the B-25 will join EAA's B-17 and Ford Tri-Motor flight experience aircraft and travel the country, bringing aviation history to life.

After suffering a main gear failure on landing, the B-25 was put on display in the museum. (Ingo Warnecke via airport-data.com)

N10V during restoration in 2019. The bomber again has an attractive colour scheme and nose art. On 20 April, 2019, "Berlin Express" made its first flight in decades after a thorough restoration. (EAA)

ERICKSON AIRCRAFT COLLECTION, MADRAS, OREGON

The beautiful restored N8195H "Heavenly Body" is currently owned by Jack Erickson at Madras, Oregon. Jack Erickson is an aviation icon. The founder of Erickson Aviation transformed the Sikorsky S-64 Skycrane into the Erickson Aircrane, a huge heavy-lift helicopter that has become a heralded platform for remote timber harvesting,

aerial firefighting, and more unconventional jobs. Erickson is also the man behind the fabulous Erickson Aircraft Collection. This collection was started by Jack Erickson in 1983. He bought a P-51D Mustang in 1980. The next airplane he bought was an F4U Corsair in 1983 and six months later he bought a unique two-seat Mark VIII Spitfire.

From then, he started looking at warbirds from a different point of view. The Erickson Aircraft Collection is located at the Madras Municipal Airport in Madras, Oregon. The original airfield was expanded during the World War II and served as training base for B-17 Flying Fortresses and Bell P-63 Kingcobras. Since the 1970s, the airport also hosts an annual Air show of the Cascades. In 1991, the collection was put on public display inside of a wooden WWII blimp hangar as Tillamook Air Museum in Tillamook, Oregon. In 2013, Erickson Aero Tankers, the business entity that owns Erickson's vintage

"Heavenly Body" at Madras Municipal Airport. (Collection Wim Nijenhuis)

N8195H "Heavenly Body" in the hangar of the Erickson Aircraft Collection. (Collection Wim Nijenhuis)

Detail of the Vargas style nose art. (Collection Wim Nijenhuis)

airplanes, moved its operations to a new facility in Madras and relocate the Aircraft Collection. Since almost all airplanes are airworthy, they were simply flown out of Tillamook in May 2014 and were displayed in a brand-new hangar at the Madras Municipal Airport. Today, the Erickson Aircraft Collection consists of about 30 mostly WWII era airplanes. The planes are regularly flown as a part of their maintenance plan and often participate in air shows around the country.

The B-25 "Heavenly Body" is generally considered as one of the best restored Mitchells. The bomber that was the first of two B-25s that launched off the aircraft carrier USS Ranger to commemorate the 50th Anniversary of the Doolittle Raid in April 1992, was bought by Erickson in January 2014. The B-25 is painted in an attractive scheme representing a ship of the 390th BS, 42nd BG "Crusaders" in the South West Pacific Area.

In 2019, the Erickson Aircraft Collection announced the launch of a nationally touring all-female WWII B-25 travelling exhibit, honouring the Women Airforce Service Pilots (WASP). This exhibit was made possible through a collaborative partnership with the Women Airforce Service Pilots Official Archive at Texas Woman's University in Denton, Texas, and the National WASP WWII Museum in Sweetwater, Texas.

The tour started on 5 and 6 October of that year at the California Capital Airshow held in Sacramento at Mather Field, a former airbase where the WASP trained on the B-25 during WWII. The bomber was provided with special WASP nose art.
(Erickson Aircraft Collection)

FAGEN FIGHTERS WWII MUSEUM, GRANITE FALLS, MINNESOTA

Located in Granite Falls, Minnesota, the Fagen Fighters WWII Museum is a fairly new museum located at the Lenzen-Roe Memorial Field. Construction of the museum began in April 2011. The museum consists of two hangars, a WWII Quonset hut, and a WWII control tower. The newest hangar and flagship of the museum is a 15,000 sq.

ft. brick tornado-proof structure. The museum is home to the pristine collection of fully operational and active WWII trainer and fighter aircraft restored and owned by Fagen Fighters. Fully operational authentic WWII ground vehicles are proudly displayed among the aircraft. A library overlooking the museum's first floor containing WWII

books and documents collected over many years is a work in progress. Authentic uniforms and artefacts are displayed throughout.

In 1974, Ron Fagen returned to his native Minnesota after a stint in Vietnam. He started a construction business with a single pickup truck and a four-man crew. Over the years, Fagen seized several opportunities to expand his business, specializing in heavy industrial construction. Today, the company that bears his name, and of which he is CEO, is a large contractor in the U.S. While the company takes on many different types

of industrial construction projects, Fagen is the largest green energy design-builder in the country, specializing in building ethanol plants and wind farms. Late 1990s, Fagen became an industrial legend during the construction boom days of the U.S. ethanol industry. Fagen Construction is still going strong in the energy world including erecting steel towers for wind farms across the north central states.

Ron Fagen and his wife Diane created Fagen Fighters World War II Museum. It was started with one airplane, a P-51 Mustang purchased in November 1994. They bought another WWII fighter a few years later, then another, and another. And because word travels fast in a small town, they were starting to get visitors and so they started a museum. The Fagan's have since bought and restored World War II airplanes from around the world. They created work for their employees during the recession by having them build a tornado-proof hangar. They officially opened their doors to the public in 2012.

One of the larger airplanes of the museum is the B-25J-30 with s/n 44-86698. She was delivered in June 1945 and was initially flown to storage in Salina, Kansas. By November 1945, she was placed into consolidated storage at Independence, Missouri and moved to Pyote, Texas in September 1947. In October 1951, she was removed from storage, overhauled in Alabama, and was transferred to the RCAF in December 1951. After nearly 10 years of service, she was removed from the RCAF inventory and the Canadian registration CF-NWU was as-

"Paper Doll" of Fagen Fighters with her beautiful paint scheme and nose art as used in the 447th BS of the 321st BG in the MTO. (Fagen Fighters)

signed in December 1961 when she was sold to Hicks & Lawrence Ltd. of Ostrander, Ontario. She came back to the U.S.A. in 1967 and was registered as N543VT. She got several U.S. owners and the bomber went to Canada again in August 1982. There, she was registered as C-GUNO and flew as a fire bomber. In 1992, she was withdrawn from service and in October of 1998, she was purchased by Aerocrafters in Santa Rosa, California. Her current registration was as-

signed as N325N. She was restored to her previous military glory and fitted with a 12-gun nose from 1998 through 2001. She flew as "Sunday Punch". In February 2012, she was bought by Fagen. She was restored by Aero Trader and fitted with a glass nose. She is now operated by Fagen Fighters and is flown in an Olive Drab scheme as "Paper Doll", remembering a B-25 of the 321st BG, 447th BS at Solenzara, Corsica in 1944.

The beautifully restored Olive Drab/Neutral Grey "Paper Doll" at Oshkosh in July 2014. (Johan Hetebrij)

Fargo Air Museum, Fargo, North Dakota

The Beck-Odegaard Wing is the home of a B-25 in the Fargo Air Museum. This museum is housed in two hangars near Hector International Airport northwest of Fargo, North Dakota. The museum's extensive collection of flying machines and paraphernalia includes military, commercial and recre-

In 2014, the N9641C was towed from Wahpeton to the Fargo Air Museum. (Meredith Mitskog)

The bomber photographed in the museum in September 2017. (Nate Nickell, Collection Wim Nijenhuis)

ational airplanes. The Fargo Air Museum was established in 2001 as an organisation that promotes interest in aviation through restoration, preservation, and education. Gerry Beck and Bob Odegaard, two late area aviation pioneers, and two others founded the museum, which opened in 2001. The Beck-Odegaard Wing is named for the two men who helped establish the air museum and both died in airplane accidents. Beck, of Wahpeton, was known for his restoration of vintage World War II-era planes. He died in 2007 at an air show in Oshkosh, Wisconsin. Similarly, Odegaard was killed during an airplane crash while practicing for an air show in Valley City, North Dakota in 2012. Fargo Air Museum began taking flight in the late 1990s, when a group of local military folks, pilots, and airplane restorers realised they were not alone in their passion for aviation. They opened the first hangar in 2001, and the museum has continued to grow ever since. Today, various airplanes are on display, including a Wright Flyer replica, Huey helicopter, and a B-25 Mitchell.

This B-25 is s/n 44-30010, a J-25 model. She was stored at Davis Monthan AFB in December 1957 and sold to Arrow Sales Co. at Hollywood, California in 1958. She was registered as N9641C. Probably around 1962, she was sold to H.H. Coffield of Rockdale Flying Service, Texas. She was stored in original USAF markings until October 1983 when she was sold at an auction to Gerry Beck of Tri-State Aviation, Wahpeton, North Dakota. Beck flew her back to Wahpeton during a February snowstorm in 1984. Once Beck and his crew got the airplane home and started looking at her, they discovered she had been far more corroded from the ocean air than they expected. He was lucky to have gotten her home, and to get her in airworthy condition, she would have taken a lot of work. So, Beck and his wife Cindy, decided the airplane would be donated to the Fargo Air Museum someday. The bomber was stored in a hangar with her original markings 0-430010/499. Beck donated the bomber to the Fargo Air Museum in 2006. In 2014, the bomber was towed to the Fargo Air Museum. The reason it has taken so long for the airplane to make it to Fargo was because the museum simply did not have the room. But with the addition of the Beck-Odegaard Wing in 2014, it was finally able to make her to where she belongs.

Flying Heritage & Combat Armor Museum, Everett, Washington

An ex RCAF B-25 is flown by the Flying Heritage & Combat Armor Museum at Everett, Washington. This is a B-25J-25 with s/n 44-30254. In Canada she was civil registered as CF-MWC and was operated as a tanker. In June 1995, she was sold to Jeff Thomas at Anchorage, Alaska, and registered as N41123. Jeff Thomas is an American Airlines 777 captain and aircraft collector. He was the son of a Navy mechanic who had worked on Wildcats and Hellcats in the Pacific. Thomas had done some professional restoring himself, and owned Vintage Wings, a historic aircraft restoration business. In February 1999, the bomber was sold to Paul Allen of the Flying Heritage Collection at Bellevue, Washington. Allen had a passion for aviation and history, and his awareness of the increasing rarity of original World War II airplanes motivated him to restore these planes and share them with the public. So, the B-25 was delivered to Aero Trader, Chino, California for restoration to military configuration and to airworthy condition. She made her first flight on 7 March, 2011. She was delivered to Paine Field on 9 June, 2011 and is flown in an Olive Drab USAAF scheme.

In 1998, Paul Allen began acquiring and preserving vintage aircraft and armour. The collection opened to the public in 2004 at the Arlington, Washington, airfield, but in 2008 moved to a newly renovated historic industrial hangar located at Paine Field in Everett, Washington. In 2013, the Flying Heritage & Combat Armor Museum added a 22,000 square foot expansion hangar for its expanding collection. In the late summer of 2018, the museum opened a third hangar, expanding its campus and adding new airplanes and tanks to the ranks. On 24 March, 2017 the Museum changed its name from the Flying Heritage Collection to the Flying Heritage & Combat Armor Museum to reflect the transition from exclusively airplanes to a military vehicle & history Museum. In the working hangar are military artefacts from the United States, Britain, Germany, Soviet Union, and Japan, acquired and restored with unparalleled authenticity to share with the public. The

Flying Heritage & Combat Armor Museum contains two dozen airplanes of the type that saw action during World War II, including German Messerschmitt's, Russia Polikar-

pov's, and American airplanes. In addition to airplanes, the collection also preserves other weapons of war.

During landing at Skyfair 2018, Paine Field.
(Gary Maisack)

The Olive Drab N41123 in 2017 at the Skyfair air show. The winged skull squadron emblem with a white border on the nose is from the 341st BG, 490th BS that flew in the China-Burma-India Theatre. The hop-bombing technique of the 490th BS became so successful that the squadron earned the nickname "Burma Bridge Busters". (Cafsocal)

The airplane is running her engines during the Flying Heritage & Combat Armor Museum Flyover at 2018 Seahawks Training Camp. (FHCAM)

Flying Tiger Air Museum, Paris, Texas

The Flying Tiger Air Museum had a collection of World War Two and Korean War era surplus warbirds that occupied for many years a small air strip west of Paris, Texas. The museum was founded in the early 1970s by I.N. Burchinal Jr. He served in the U. S. Coast Guard and was an accomplished pilot. At the Flying Tiger Air Museum, he collected, restored, and taught flying lessons in vintage World War II warbirds. He had flown as a stunt pilot for Universal Studios, and his planes were featured in several movies including "The Great Waldo Pepper", "Baa, Baa Black Sheep" and "Midway". Burchinal provided training and checkout rides for most of the types on the field tailored closely to the student's skill level. Pilots with 100 hours or 10,000 hours often showed up for a quick instructional ride in the aircraft of their choice. Burchinal worked for the telephone company for years, crop dusting in the early mornings and late evenings, fooling around with airplanes every spare minute. Eventually, he left the phone company, set up his own airport and started dusting and instructing full time. Working out of his Northeast Texas crop dusting facility, he collected leftover military airplanes and fulfilled dreams for anyone wanting to learn to fly them. He bought his first fighter, a P-40, for $300 in the early 1950s, and since then has owned just about everything the military ever operated. Unfortunately, legal problems in the late 1970s/early 1980s forced Burchinal to sell off most of the collection. Many of his former aircraft are now in various museums scattered across the country. There isn't much left of the old museum. Today, just an old Martin 404 passenger airplane resting out front of an old hangar, a crippled old Piper Cub, and a few old Cessna fuselages discarded out back in the weeds.

Registration records indicate Burchinal at one time or another had seven B-25s registered in his name: N96GC, N543VT, N3680G, N3698G, N9115Z, N9446Z and N9899C.

N96GC

The B-25 used by Grand Central Aircraft Co. was purchased from Cal Aero Technical Institute, Glendale in August 1950. She was registered as N67998. By July 1951, this B-25H was converted for civil use and fitted with a glass nose. She was modified with a B-25C tail stinger and a passenger interior. In June 1957, the glass nose was faired over,

Not the best picture, but nevertheless a nice view of Burchinal's N543VT in flight. On both sides of the nose she had a nose art with the text "Ruptured Duck", probably referring to one of the Doolittle Raiders in April 1942. (Budd Davisson)

N543VT was s/n 44-86698. She is shown in the 1970s at the Texas airfield of Isaac Burchinal. In January 1969, she was purchased by Burchinal for display at his Flying Tiger Air Museum and use for its flight training school. (Malcolm Taylor)

N3680G was for a short time owned by Burchinal. In July 1976, she was purchased and later that year, she was already sold again. (Collection Wim Nijenhuis)

The overall aluminium finished N3698G is seen here at Opa Locka, Florida in the early 1970s. *(Larry Johnson)*

Below: N3698G in the early 1980s, after she was sold by Burchinal. Her new owner Donald Webber of Baton Rouge, Louisiana named the airplane "Cochise". *(Collection Wim Nijenhuis)*

and the waist windows were modified. In July 1957, she was registered as N96GC. She got some different owners and in 1973, she was sold to Burchinal. In 1974, she was sold to Charles V. Moody, Tampa, Florida and later to William J. Parker, Atlanta, Georgia. The airplane crashed in June 1975.

N543VT

This was a former Canadian airplane and registered as CF-NWU. She came back to the U.S.A. in 1967 and was registered as N543VT. In January 1969, she was purchased by Burchinal for display at his Flying Tiger Air Museum and use for its flight training school. In September 1975, she was sold to Daniel Jackson, Seymour, Texas who loaned her to the SST Aviation Museum in Kissimmee, Florida.

N3680G

A B-25J-35 with s/n 45-8887. In August 1957, the bomber was sold to H.H. Coffield, Rockdale, Texas and registered as N3680G. She was acquired from USAF disposal and converted to executive configuration which was completed in January 1960. In October 1970, she was sold to Rockdale Flying Service, Rockdale, Texas and in December 1972, to Central American Imports, Phoenix, Arizona. In July 1976, she was purchased by I.N.

N9446Z was purchased by Burchinal in December 1972. Here she is seen at Gillespie Field in El Cajon, California a year before in 1971. She crashed on 6 August, 1976 in Chicago, killing both pilots. *(Dusty Carter)*

Burchinal. Later that year, she was already sold again to North American Transport Co., Tucson, Arizona. In December 1976, she was damaged by an undercarriage collapse during landing at Vega, Texas. Currently, she is stored in parts at Mitchell Aircraft Components of Carl Scholl at Chino, California.

N3698G

In September 1974, the then owner of this B-25J-20 s/n 44-29507, Ernest G. Trapaga from Redondo Beach, California, sold the bomber to Burchinal. Five years later in September 1979, she was sold to Robert Wingate, then back to Burchinal in November, who sold her to Reyline Aviation from Kissimmee, Florida. In June 1981, she was sold to Donald Webber of Baton Rouge, Louisiana where she flew as "Cochise". Currently, this airplane is owned by the Royal Netherlands Air Force Historical Flight.

N9115Z

This ship, a B-25J-20 s/n 44-29366, was used by Sonora Flying Service, Columbia, Cali-

This is N9899C at Amon Carter Field, Fort Worth, Texas in April 1976, shortly before she was purchased by Burchinal in August of that year. *(Carl Jenkins)*

fornia as fire tanker #48. The airplane was registered as N9115Z and sold to Sam Rawland & Morgan Hetrick, Osage, Missouri in October 1964. In April 1968, she was sold to Burchinal. One year later, the ship was sold to Filmways Inc, Hollywood, California and flew in the movie "Catch-22". Later she was used for filming "Hanover Street" in the U.K.

N9446Z

This B-25J-25, s/n 44-30737, was registered

N9899C at Paris, Texas. Unfortunately, on the picture the nose is cropped. She was a sister ship to N543VT, as it carries the same nose art. The B-25 went to Isaac Burchinal, in whose care she was photographed in the 1970s at the Flying Tiger Air Museum. The Paris Museum is now long gone. (Malcolm Taylor, Mark Fidler)

FREEDOM MUSEUM USA, PAMPA, TEXAS

as N9446Z. She was purchased by Burchinal in December 1972. In July 1974, she was sold to Air Chicago Freight Lines, Chicago, Illinois. The airplane was destroyed in a fatal crash in Chicago on 6 August, 1976 when she was in service with War Aero Inc. from Chicago. Apparent cause was engine failure during take-off.

N9899C

This B-25 J-20, s/n 44-29127, was registered as N9899C in September 1958. She was purchased by Burchinal in August 1976. After eight years, she was sold to Colvin Aircraft Inc. at Big Cabin, Oklahoma in September 1984. There were some major disputes over this airplane. There is no indication in the FAA registration records that she was involved in a belly landing. This airplane is now reported stored in Anoka, Minnesota.

Here is the ex–Bolivian ship displayed at Quantico as a PBJ–1D in three tone blue Marines scheme. The museum initially opened in May, 1978, and was closed in November 2002. The airplane then went to the Freedom Museum USA at Pampa.

(Collection Wim Nijenhuis

The Freedom Museum USA is one of the premier military museums in the Texas Panhandle and located in Pampa, Texas. The museum has one B-25 in her collection. This ship is a B-25D-30 and was originally registered as N8011. In 1966, she was exported to Panama and in 1970, exported to Bolivia and registered as CP-915. About 1976, she was withdrawn from use and was left derelict at La Paz. She returned to the United States in 1987 to Roy M. Stafford from Jacksonville, Florida. Mike Parker, President of the B-25 Preservation Group, was able to get a lead on the B-25 for the museum from Quantico and the B-25 was transferred to the Group. She was restored to static airplane and displayed as USMC PBJ-1D at the USMC Air-Ground Museum in Quantico, Virginia. But the museum closed in 2002. She is now on display at the Freedom Museum USA in Pampa, Texas.

Currently, the PBJ-1D is displayed outside at the Freedom Museum USA. More than 65 years of age difference. In April 2016, children visited the museum and played under the tail of the old Mitchell. (Pampa Freedom)

The Freedom Museum USA was opened in 1994. In August 1986, some Pampa Army Airfield people were talking and suggested they might establish a museum. An old pump station built in 1939 was already a historical building and needed tender loving care. The city approved and called for architectural drawings. Chuck Dempsey started the Pampa Army Airfield Museum with a gift of Wilbur and Orville Wright artefacts. A merger was accomplished with the Veterans of Foreign Wars, and funds were raised. Southwest Industries got the contract and built the building. After eight years, the doors were opened on 20 August, 1994. Housed in the old water pumping station built by the Public Works Administration, Pampa's Freedom Museum USA features a large collection of military memorabilia, featuring displays provided by the Army, Air Force and Marines, indoors and out.

June 2015, on approach into McNary Field, also known as Salem Municipal Airport, Oregon. (7XDriver)

HISTORIC FLIGHT FOUNDATION, PAINE FIELD, WASHINGTON

The Historic Flight Foundation has a small but excellently preserved collection of airplanes, with an airworthy B-25 named "Grumpy". The Historic Flight Foundation has choosing to focus on the period 1927 - 1957, from the dawn of civil aviation through to the start of the commercial jet age and features both civil as well as military aviation. The museum is located at Paine Field, Washington, and end 2019, some of the collections are displayed in a new facility at Fets Field, Spokane, Washington.

They currently have eight airplanes in their collection, including a P-51 Mustang, Spitfire and the mentioned B-25. The Foundation offers the opportunity for members of the public to fly in their planes. Beginning in 2006, the primary focus became planning for a permanent facility to share the collection, maintain the aircraft for flight, and restore aircraft. The Restoration Centre

opened on March 2010, allowing for the collection to be conveniently presented to the public. The Restoration Centre is presented as 'a working hangar'. They also have a second hangar in which the 'dirty' mechanical work is done, and they have ambitious plans to build a new Education Centre adjacent to the current hangar.

Historic Flight Foundation was established in 2003 as "John T. Sessions Historic Aircraft Foundation" with the intention to collect, restore and share significant aircraft from the period between the solo Atlantic crossing of Charles Lindbergh and trans-Atlantic service in the Boeing 707. John T. Sessions was a Seattle area attorney, businessman, entrepreneur, and philanthropist. He traces his interest in aviation back at least as far as when he was a young attorney on staff at Boeing and has parlayed his own growing interest in aviation into a personal collec-

In 2015, "Grumpy" received a new paint carried out by Sealand Aviation in Campbell River, British Columbia, Canada. After completion of the paint job that took some months, the bomber returned to Paine Field. (Sealand Aviation)

Right: *May 2019, wing flaps down, "Grumpy" arrives at Paine Field with the new Paine Field passenger terminal and its control tower in the background.*
(Kent Mathiessen)

tion and now into this public museum. Sessions has also become a broadly qualified pilot and flies many of the planes the Foundation now owns.

The B-25 of the Foundation is a B-25D-30 model and registered as N88972. This airplane was displayed and flown for several years with The Fighter Collection at Duxford in the United Kingdom. She was painted with the RAF serial KL161 and aircraft code VO-B of No. 98 Squadron and named "Grumpy". In 2003, she was placed in stor-

age at North Weald, U.K. In 2008, she was purchased by Historic Flight Foundation and restored to airworthy condition by Aircraft Restoration Company in Duxford, England. In 2009, she returned to the U.S.A. She is still flying in her RAF colours as "Grumpy".

KANSAS CITY WARBIRDS INC., KANSAS CITY, MISSOURI

The first nose art of the plane was applied in 1986 and above, the second nose art on the ship was painted in 1987 by aviation artist Kevin Lundy. (Robert Bourlier, Jim Stella)

A total of 6.680 B-25s were produced at the North American Aviation plant in Kansas City. Reason for a group of people to create a living museum as a tribute to and remembrance of all the bomber builders who had worked at the plant and its significance for the country and the local environment. The latter is best known under the name Fairfax District. So, this group of people formed the Kansas City Warbirds Inc. which was incorporated on 24 May, 1977. Their aim was to acquire and operate military aircraft. In addition to a management, the corporation also had a membership. At various times, they had a number of individual stockholders, but investors gradually withdrew, and Frank Howerton and Barney Pool became the primary owners. In early 1977, the group had planned to bring back a B-25 to Kansas City. They located a B-25 at Chino, California that was owned by John Stokes at San Marcos, Texas. So, on 17 May, 1977 the B-25 was purchased. It was civil registered as N6123C. One month later 13 June, she flew back to Kansas City. The airplane was in a rough shape when she returned to Kansas City after several years of storage in the Arizona desert. They started an extensive restoration. They teared down and rebuilt the

whole airplane. After years of restoration the bomber had 10 seats. After restoration, she made her first test flight on 29 February, 1984. The airplane was overall natural aluminium finished and had a white tail. During the war, the Modification Centre of the North American plant modified B-25Ds into white tailed F-10's photoreconnaissance airplanes. Therefore, they wanted the B-25 after the restoration in this configuration. The plane was named "Fairfax Ghost". She was built at Fairfax and had returned there, so it was an appropriate name for the bomber. In September 1986, a nose art was painted on the fuselage nose but replaced with a new one in the summer of 1987. After the bomber was restored, she became a frequent participant in air shows. But the Kansas City Warbirds Inc. had a chronic financial underfunding. So, in June 1993, the "Fairfax Ghost" was flown to Rockford, Illinois for a broker to sell. With this, Kansas City Warbirds Inc. came to an end and in September 1993, the corporation was legally dissolved. The airplane was bought in 1994 by The Flying Bulls from Austria. Nowadays, she is part of the fleet of historic aircraft of the "The Flying Bulls" and often participant in European air shows.

She was participant in air shows as in these pictures taken in 1988 and in 1993. (Collection Wim Nijenhuis)

LAURIDSEN AVIATION MUSEUM, BUCKEYE, ARIZONA

The Lauridsen Aviation Museum at Buckeye, Arizona, was founded in 2012 by Hans Lauridsen. He has a private collection of over 30 aircraft, including a B-25. He also has a A-26, PBY-5A, DC-3, C-45, C-119 and many others. The museum could be described as a museum in the making. Most of the collection of

The museum is looking for a location that offers room to display existing aircraft and also has space available for future growth. The B-25 of Lauridsen is N3438G. In June 2004, she was purchased by Hans Lauridsen from Wiley Sanders Truck Lines in Troy, Alabama, where she flew as "Samantha" and

LEWIS AIR LEGENDS, SAN ANTONIO, TEXAS

Lewis Air Legends is not a museum open to the public. It is a flying museum that goes to the public. It has a great collection of airplanes started with one T-28 Trojan trainer Rod Lewis purchased in 1995. Lewis Air Legends is the private collection of World

"Ol Gray Mare", here photographed in April 2007 and owned by Hans Lauridsen.
(Scott Slingsby)

N3438G in November 2011. The name and nose art "Ol Gray Mare" have disappeared and have been replaced by "Auntie Jayne". By 2014, this last name had also disappeared. (AFIA, Alan Wilson)

airplanes is located at the Buckeye Municipal Airport, west of Phoenix, Arizona. However, some of the airplanes are stored at several other nearby airports. Hans Lauridsen is planning to erect a 30,000 sq. ft building to exhibit more valuable planes inside while others will be parked on the surrounding tarmac. His vision is to allow the public to get close to the aircraft without any ropes and see them being worked on with engine cowlings open.

later as "Ol Gray Mare". Around 2011, Lauridsen replaced the name and nose art with "Auntie Jayne". A few years later, this name was removed again, and the bomber was displayed without any nose art. In 2020, she was set up outside in the open air and her condition did not seem to improve. Therefore, it is hoped that Lauridsen soon finds a new location where a building can be erected that can accommodate the B-25 and the other warbirds of the museum.

War II and Korean War fighter and bomber aircraft owned by Rod Lewis, principal of Lewis Energy Group. Over the past 35 years, Rod Lewis, founder and Chief Executive Officer of Lewis Energy Group, has climbed his way to the top of the international oil and gas industry. Lewis bought his first well in 1982, and the oil and gas empire that is now Lewis Energy Group was born. As son of an Air Force pilot, Lewis was an amateur airplane mechanic

long before he was an oilman. From his first plane, his collection grew to include over 24 classic WWII airplanes. The collection includes some of the most significant and best-restored aircraft of this era, including a B-25 Mitchell bomber. The aircraft appear at air shows in Texas and the south-western United States

Lewis personally flies every one of his planes, including the vintage warbird collection known today as the Lewis Air Legends.

The Mitchell of Lewis Air Legends is a B-25J-25, s/n 44-30456. She was delivered in January 1945. Accepted surplus to military needs, she was initially stored and in the 1950s used as a trainer. After storage at Davis Monthan, she was sold to Skyways System of Miami, Florida in July 1959, and her registration of N3512G was reserved. In the 1960s, she was sold several times and modified with a portside waist cargo door and was sporadically used to transport cargo. In May 1982, she was sold to William Arnot of Breckenridge, Texas and registered as N43BA. She was restored to airworthy condition and flew as "Silver Lady". During that restoration, the cargo modification was removed, and the aircraft was restored to original configuration. In August 1994, she was sold to Jack Erickson of Central Point, Oregon. In 1997, she was purchased by Tillamook Air Museum, Tillamook, Oregon and displayed there until 2007. In 2008, she was purchased by Lewis Air Legends and

restored to airworthy condition. Her current registration of N747AF was issued and she made her first flight after restoration in March 2010. Aero Trader at Chino, California, completed the original restoration of the Lewis Air Legends B-25. The restoration took a total of 18 months and included a complete interior overhaul with new instru-

ments and interior refurbishment, along with the new paint scheme, bringing the plane to excellent condition. She is painted in Soviet colours and features a blonde gal decked out in leather riding on a bomb with the name "Russian Ta Get Ya" painted by pin-up artist Greg Hildebrandt in 2011.

LIBERTY AVIATION MUSEUM, PORT CLINTON, OHIO

Exactly 50 years after the Japanese attack at Pearl Harbor, the Liberty Aviation Museum was established on 7 December, 1991 by a handful of volunteers. In 1994, they held an air show at the Erie-Ottawa Regional Airport in Port Clinton, Ohio and this led to an announcement in 1996 about their intention to locate a museum at the Erie-Ottawa Regional Airport. They planned a facility including a hangar and museum along with a 1950's era themed Diner. They already possessed a Ford Tri-motor and the restoration of this airplane led to the establishment of the Tri-motor Heritage Foundation. In 2011, a B-25 was acquired, and the 2011 season was spent training crews to fly this bomber at air shows throughout the year.

The same year, plans were announced for a permanent museum/hangar facility at the Erie-Ottawa County Regional Airport. The museum facility opened to the public in July 2012 and was heralded by the arrival of the newly refurbished B-25, "Georgie's Gal". In 2015, a second main hangar, along with an attached workshop, a multi-purpose hangar and a three-story WWII themed control tower, attached to the new hangar was completed and opened in 2015.

The museum's B-25 "Georgie's Gal" started her life at the North American plant in Kansas City, being accepted into USAAF service in 1945 and was used for both pilot and navigator training. She was retired from service in August 1958 and stored at Davis Monthan. In January 1959, she got her civil registration N9167Z. She changed ownership several times. In the 1970s, she was operated as "The Devil Made Me Do It". By 1992, her registration was changed to N345BG and she flew later as "Martha Jean". The Liberty Aviation Museum purchased the plane in 2011, and from winter 2011 through spring 2012, she went through an extensive mechanical and cosmetic restoration at Aero Trader in Chino, California. Her first flight after restoration was in July 2012. She was renamed "Georgie's Gal" in honour of the museum's main patron George V. Woodling Jr., a Cleveland-area history and aviation buff who died in 2010. The air-

The bomber of Lewis Air Legends in 2014 with the registration N747AF. She has a three-tone Soviet colour scheme and a beautiful nose art. (Valder137, Adrian Share)

The nose was made by artist Greg Hilde-brandt in 2011. This is the stunning result.
(Collection Wim Nijenhuis)

plane had a completely new paint scheme and nose art. The B-25 can now be seen at air shows around the central and eastern United States.

Lone Star Flight Museum, Houston, Texas

The Lone Star Flight Museum is located at Ellington Airport, Houston, Texas. The museum relocated from its previous Galveston, Texas location to Ellington Airport in September 2017. The vintage military aircraft museum in Galveston was swamped by Hurricane Ike in 2008 and survived Harvey after moving to Houston. Opened in 1987 and located at Scholes Field in Galveston, the museum began with the private airplane collection of Robert L. Waltrip. In 1984, he purchased a B-25 Mitchell and began collecting and restoring many of the World War

II aircraft he admired as a boy. He formed the Lone Star Flight Museum in 1987 to share what would become one of the finest collections of historic airplanes in the world. By 1990, the collection had grown so significantly that the owner decided to place them on public display and convert the collection into a non-profit organisation. The museum boasts a fine aviation collection. In addition to a collection of award-winning historic airplanes, the 130,000 square-foot museum features interactive, educational exhibits focusing on the science, technol-

A beautiful nose art is painted on both sides of the fuselage nose. (Wim Nijenhuis)

Ex "Martha Jean" in 2017 at Grimes Field, Urbana. She has an Olive Drab/Neutral Grey camouflage scheme and is named "Georgie's Gal". (Dennis Nijenhuis)

ogy, engineering, and math concepts essential to flight while allowing visitors to explore Texas' rich aviation heritage. Another highlight of the new facility is an Aviation Learning Centre and the Texas Aviation Hall of Fame. With more than 40 displays of significant flying vehicles and many hundreds of artefacts related to the history of flight, the collection honours the contributions of native Texans and residents including Howard Hughes and former President George H. W. Bush through its Texas Aviation Hall of Fame. Sheltered in the museum's hangars, an array of historically significant aircraft illustrates the technological innovations discussed in the galleries. Most of the

Lone Star Flight Museum's airplanes date to WWII, while some served in later conflicts. In addition to displaying many aircraft from its permanent collection, the museum hosts a rotation of planes from other organisations and individuals.

One of the most important airplanes of the museum is a B-25J-30, s/n 44-86734. This ship was delivered in June 1945 and in the late 1940s and 1950s, she was mainly used for administrative duties. In October 1958, she was flown to storage. In December 1959, she was sold to American Compressed Steel Corp. of Cincinnati, Ohio and registered as N9090Z. In December 1960, she

was sold to Aero American Corp. of Cincinnati, Ohio. Between July 1961 and February 1979, she was sold several times, and, in this month, she was sold to Dean Martin of East Middlebury, Vermont. Her registration was changed to N600DM. In August 1984, she was sold to Robert Waltrip and her registration was changed to the current N333RW in December 1985. She was beautifully restored and flew in white and blue colours as a PBJ-1 named "Special Delivery". In April 2007, on the occasion of the 65th Anniversary of Doolittle's Tokyo Raid, the museum made the decision to change the livery of their Mitchell from USN colours to those of Doolittle. She is the only civilian airplane to feature the Doolittle Raider emblem flying as "Doolittle Raiders Special Delivery".

The white and blue PBJ-1 "Special Delivery" of the Lone Star Flight Museum. The nose art on both sides of the airplane was found on many airplanes during World War Two. The painting is based on Alberto Vargas' flying woman wearing a one-piece swimsuit and a long, flowing scarf. It was titled "There'll Always Be a Christmas" and appeared in "Esquire" magazine in December 1943. (Christopher Ebdon, Collection Wim Nijenhuis)

LYON AIR MUSEUM, SANTA ANA, CALIFORNIA

Nowadays, a beautiful highly polished B-25 is the B-25J-20 with s/n 44-29465, at the Lyon Air Museum in Santa Ana, California. The museum was founded by Major General William Lyon and is located in a hangar on the west side of the John Wayne Airport in Orange County, California. The Lyon Air Museum contains many historic aircraft, rare automobiles, military vehicles, military motorcycles, and other memorabilia related to World War II. General Lyon also was successful in business, having founded Lyon Homes, large homebuilders. The museum is a facility with tours led by volunteer docents and interactive educational exhibits intended to give visitors historical narratives about the history of each aircraft. During his distinguished military career, Major General Lyon held the position of Chief of the U.S. Air Force Reserve from 1975 to 1979. He served

in Europe, the Pacific Region and North Africa during World War II and later flew combat missions in Korea. His superbly conditioned and operational private collection of historically significant aircraft includes a Boeing B-17, Cessna O-1E, Douglas DC-3, Douglas

N25GL named "Guardian of Freedom" owned by Glenn H. Lamont of Detroit, in August 1994 at the Tillsonburg Air Show, Ontario.

(Mike Henniger)

C-47, B-25, and a Douglas A-26. A small but impressive collection in a sparkling clean and mint condition. In addition, the museum displays military vehicles and motorcycles, vintage automobiles, and related period memorabilia. General Lyon's passion

In the hands of the Lyon Air Museum, she received another design of the nose art.

(Richard Vandervord)

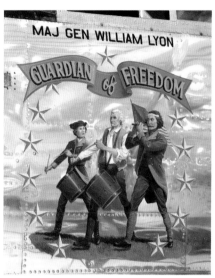

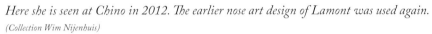
Here she is seen at Chino in 2012. The earlier nose art design of Lamont was used again.
(Collection Wim Nijenhuis)

for aviation history and youth education is the driving force behind the museum, for which ground-breaking took place in 2007. The Lyon Air Museum officially opened the doors to its new 30,000-sq.ft. facility at John Wayne Airport in December 2009. The sleek building is a hybrid of metal building systems and conventional construction, and shares its location with Martin Aviation, an Air Lyon, Inc. company also owned by the general. It incorporates three Martin Aviation hangars previously at the site, which provide maintenance for corporate aircraft and general aviation at the airport.

The B-25 bomber of Lyon Air Museum was delivered in September 1944 to the USAAF. During and after the war, she was used for pilot and technical crew training. In December 1958, she was stored at Davis Monthan. In July of the following year, she was sold to Haddock Motor Sales in Battle Creek, Michigan. In October 1959, she was sold to Ernest Beckman and William Haddock at Battle Creek and registered as N3523G. From 1965 to 1975, she was withdrawn from use and placed in open storage at Battle Creek. In July 1975, she was sold to Glenn H. Lamont of Detroit, Michigan. She was ferried to Detroit in 1977 and was restored. In December 1985, her current registration of N25GL was assigned. By 1989, she was airworthy and was flown as "Guardian of Freedom". In June 2000, she was sold to the Lyon Air Museum.

MARCH FIELD AIR MUSEUM, RIVERSIDE, CALIFORNIA

The March Field Air Museum is an aviation museum at Riverside, California, adjacent to March Air Reserve Base. The museum was founded in 1979 as March Air Force Base Museum. It moved to a new location at the base's former commissary, where it reopened to the public in 1981. In 1993, the museum moved to its current site. In 1996, as the U.S. realigned its military bases and March became an Air Reserve Base, responsibility for ownership, management and operation of the museum was turned over to the March Field Museum Foundation. In the years since, the museum has added buildings, collections, and staff members, and worked hard to appeal to a diverse audience of families, veterans, and military/aviation enthusiasts. The March Field Air Museum is a private, non-profit education organisation. The March Field Air Museum today displays large collections of military airplanes including a B-25. The B-25

N3174G after restoration with "The Crusaders" markings of the 42nd BG. Nowadays, the name "Problem Child" and nose art have been removed.
(Shamu28, Richard Bauman)

at March Field is s/n 44-31032 and was registered as N3174G in 1958. She flew in the movie "Catch-22" as "Free, Fast, Ready". In the 1970s, she was stored. The museum received the airplane by truck in 1982 and she is on loan from the USAF. She is displayed outdoors and was named "Blonde Bomber". She was restored in 1998, wearing colours of the 42nd Bomb Group "The Crusaders" and was named "Problem Child". Recently, the name and nose art have been removed.

N3174G "Blonde Bomber" at the March Field Air Museum shortly before she was restored in 1998. (emd)

MID-AMERICA AIR MUSEUM, LIBERAL, KANSAS

A nice B-25 with markings of the colourful "Air Apaches" Bomb Group is displayed in the Mid-America Air Museum. This museum is the premier aviation museum in Kansas, and it has the fifth largest collection of military and civil airplanes in the United States. It is home to over 100 airplanes and is located in Liberal, Kansas, at the Liberal Mid-America Regional Airport. Liberal was once home to a B-24 Liberator pilot training base during World War II. Liberal Army Airfield was active for three years, and then it became the Liberal Airfield and eventually the Mid-America Regional Airport. Tom Thomas and a group of local citizens founded the Mid-America Air Group in 1987. The concept was the result of a reunion in 1985 of bomber group personnel who had been stationed at Liberal during World War II. A large portion of the museum's success can be directly attributed to Tom Thomas. Over

a period of years Colonel Tom A. Thomas, Jr. USAF (ret.), had built a remarkable collection of many WWII antique and vintage airplanes. Thomas formed the flying air museum, the Mid-America Air Group, and took

the planes to air shows and airports around the country where pilots performed an air circus for audiences of all ages. Thomas eventually housed the aircraft collection in Liberal, where volunteers were working to create an air museum in a former Beechcraft plant. Thomas loaned the museum his collection for 10 years, then in 1997, he donated 53 airplanes to the collection, at the time a $3 million value. That same year the city of Liberal took responsibility for the museum and changed the name to the

May 1979, the new owner Ray Cunningham flew N9462Z for the first time to Lawton, prior to its restoration into a warbird. (Charles Duggar)

October 1981, Ray Cunningham starts the engines of the restored airplane at the Lawton Municipal Airport tarmac. The bomber is painted in the colours of a ship of the 498th Bomb Squadron, 345th Bomb Group in the South West Pacific Area. (Charles Duggar)

After Tom Thomas purchased the airplane, the colour of the nose art was changed. The red colour, however, was never carried by the 498th BS. Here she is seen in August 1987 at an air show at Slaton, Texas. (Tom Tessier)

ly and registered as N9462Z. She was converted for use in agricultural spraying. She was sold again in 1975, 1976, and March 1979 to Kenneth R. Cunningham at Lawton, Oklahoma. He flew the B-25 as "Iron Laiden Maiden". In February 1984, the airplane was sold to Tom Thomas of the Ada Aircraft Museum in Oklahoma City, Oklahoma. Thomas also flew the bomber with the same name but another nose colour. Her last known flight was in 1988 and then she was loaned to the Mid-America Air Museum. Currently, she is still on display in the museum.

The nose art of the B-25. From 1988, she is on display inside the museum. (Steve Hall)

Mid-America Air Museum. Therefore, the museum is owned and operated by the city of Liberal. The museum continued to grow, and the types of airplanes include military, homebuilt, hang gliders and ultra-lights situated around interactive displays, including the Korean War and the 1942 raid on Tokyo.

The B-25 (s/n 44-30535) of the museum with the "Air Apaches" markings, is "Iron Laiden Maiden". She was delivered in January 1945. After the war, she was used as a trainer and by August 1958, she was flown to Davis Monthan for storage. In August 1960, she was sold to National Metals of Phoenix, Arizona, and Dothan Aviation Corp. respective-

MID AMERICA FLIGHT MUSEUM, MOUNT PLEASANT, TEXAS

This museum differentiates itself from many other museums in that the great majority of the airplanes in the museum fly on a regular basis. Mid America Flight Museum is the home for historical air tours. It is dedicated to keeping a fleet of historically important aircraft in flying condition. They fly often and they love to take veterans flying in aircraft such as the C-47 that participated in the D-Day invasion. The museum is near

Mount Pleasant Regional Airport, on the city's south side. Owner and operator of the museum is Scott Glover. He is also the founder and CEO of Mid America Pet Food. Since 1996, Scott Glover has collected military and non-military airplanes, which now form the Mid-America Flight Museum. Chartered in 2013, the museum's collection consists of airplanes built between the years 1925 and 1955. Since receiving its charter back in the fall of 2013, the museum has given many free airplane rides to veterans and their families. The museum's warbird collection includes a B-25 Mitchell,

Jim Terry owned and operated N1042B, with the airplane based at the Vintage Airplane Museum at Meacham Field, Fort Worth, Texas. The airplane flew as "Pacific Prowler", here on finals before the runway as part of the Media Day activities for the Ottawa Air Show, Ontario in June 2006. The name is written in blue letters. (John Davies)

Later, the name was written in yellow letters. (Dave Budd)

P-51 Mustang, and a Navy and Marine Corps F4U Corsair fighter. There even is a C-47 Skytrain cargo and troop carrier transport. The museum is housed in six hangars, three of which are rented from the city.

The museum's B-25, named "God and Country", went into USAAF service in March 1945 but was never deployed overseas. In 1947, she was dispatched to Scott Field in Illinois for use as a trainer. By September 1958, she was declared surplus. She was purchased by Wenatchee Air Services of Yakima, Washington and registered as N1042B. In 1962, she was sold to Tallmantz Aviation of Santa Ana, California and converted to a B-25 Camera Ship. In 1989, she was used in the United Kingdom for filming of the movie "The Memphis Belle". In 1996, the airplane

A B-25 that has travelled to hundreds of air shows and aircraft displays, as a part of a museum's educational outreach program is "Briefing Time". She has been honoured with several awards for her quality of restoration, including the Experimental Aircraft Association's "Best Restored Bomber" award in 1982. This is the B-25 Mitchell owned by the Mid-Atlantic Air Museum at Reading, Pennsylvania. This museum tells the story of men and women, manufacturers, inventors, pioneers of commercial aviation and aviation heroes from the first manned balloon

its kind in the nation and the most popular event at the Mid-Atlantic Air Museum. The Mid-Atlantic Air Museum is a membership supported museum and aircraft restoration facility located at Reading Regional Airport. The museum, founded by Russ Strine and established in 1980, collects and actively restores historic war planes and classic airliners as well as rare civil and military aircraft. In January 1988, the museum relocated its complete operations to Reading Regional Airport. Here, the museum had obtained use of a hangar, offices and airplane ramps for the continued display and operation of its growing airplane collection while actively pursuing construction of a permanent museum facility. Many of the museum's historic aircraft are often seen in the air show circuit. The museum's bomber "Briefing Time" is a B-25 with the civil registration N9456Z. In the 1960s, Paul Mantz owned N9456Z. In 1969, she was used by Tallmantz in the film "Catch-22". In January 1971, after filming, she was sold to Donald Buchele of Columbus Station, Ohio. In August 1978, she was sold to F. Gene Fisher of Boiling Springs, Pennsylvania and in 1981 she was donated to the Mid-Atlantic Air Museum and was restored as "Briefing Time'". This was a B-25 that had operated with the 489th Bomb Squadron of the 340th Bomb Group in the Mediterranean Theatre of Operations. The restored B-25 was one of the first B-25s to feature accurate interior, full armament, and a meticulously detailed paint scheme, although the grey undersides should be natural metal finished. This B-25 has a very high level of completeness and authenticity. Making this achievement unique is that all the work was done by the staff and volunteers of the Mid-Atlantic Air Museum. Many of the original aircraft parts, which were no longer available, were fabricated in the museum workshop. The restoration is complete with the famous Norden bombsight, as well as a working bomb bay loaded with six rare, real

"God and Country" of the Mid America Flight Museum at Grimes Field, Urbana in April 2017. (Dennis Nijenhuis).

went back to the U.S.A. after she was sold to World Jet Inc. at Fort Lauderdale, Florida. There she was overhauled by Tom Reilly. It required a fair amount of work by Tom Reilly to go through the airplane and deal with some corrosion issues, but the airplane came up for sale once again, now without camera nose and in a polished natural metal finish. Initially flown as "Girls Rule", she was renamed "Top Secret" by the early 2000s. At this point the airplane sported Air Corps tail stripes and markings. In late 2002, she was purchased by Jim Terry of Cleburne, Texas where she flew as "Pacific Prowler". In the fall of 2013, she was sold to Mid America Flight Museum and is now flown as "God and Country".

Another picture taken at Grimes Field in April 2017. (Wim Nijenhuis)

flight in Philadelphia in the 18th century, to the space age. Interpretation reflects not only the history of the object itself, but also the evolution of technology and the cultural environment in which it was used. The annual World War II Weekend, The Gathering of the Warbirds, is the largest of

250-pound bombs. The bomber sports a complete set of original WW II radio equipment and even a rarely seen, full suite of armour plate, which along with the 1500 pounds of iron bombs on its racks makes "Briefing Time" the heaviest B-25 flying in the world today.

Close up of the nose art on the restored airplane and the original B-25 with s/n 43-27638, of the 489th BS, 340th BG in the Mediterranean Theatre of Operations. On the original airplane, the name "Briefing Time" has been rubbed out on the fuselage side.

(Collection Wim Nijenhuis, www.warwingsart. com/12thAirForce/planes5.html via Jeff Wolford)

N9456Z "Briefing Time" of the Mid-Atlantic Air Museum has a very high level of completeness and authenticity. (Richard Vandervord)

MILESTONES OF FLIGHT MUSEUM, LANCASTER, CALIFORNIA

The Milestones of Flight Air Museum was located on the South-East end of Fox Field in the city of Lancaster, California. This was a privately owned collection run under ownership of Frank Roberts, Barbara Lilley and Mac Mendoza. It was set up in the 1970s by Lee Embree and then called the Antelope Valley Air Museum. The name Milestones of Flight Museum was adopted in 1986. Soon, the display space was limited for the big amount of material. So, in the early 1990s, a deal was made with Kermit Weeks who acquired the A-20 Havoc from the museum in return for construction of a display hangar.

It had a nice collection of about 20 aircraft under which an Armstrong-Whitworth AW.660 "Argosy", Boeing KC-97G "Strato-freighter", Douglas DC-3C, Fairchild C-119C "Flying Boxcar" and a B-25. There were two hangars filled with both vintage and modern aircraft, along with various model aircraft, drawings and paintings and posters covering flight from World War II until the conquest of space flight. Unfortunately, by 2015, the museum was closed, and the aircraft were being disposed of.

The B-25 is a very rare B-25C. Her registration of N3968C was assigned in July 1952. The airplane was sold to Hughes Tool Company in January 1953 and was converted into an executive transport for the company. In July 1974, she ended up in a bad condition in Lancaster with Antelope Valley Aero Museum. She was partially restored and got new paint, but neither this museum nor its successor, Milestones of Flight Museum, did complete the restoration of the B-25. She was stored in the open field

N3968C was converted into an executive transport for the Hughes Tool Company. After her service with Hughes, she was stored in the open field at William J. Fox Field of the Milestones of Flight Museum. This picture was taken in September 1988. She is a bad condition and needs to be restored. (Peter Garwood)

without her outer wings as a derelict at William J. Fox Field through the 1980s. After the museum closed in 2015, the airplane was sold in August 2015, to San Simeon Air LLC of San Francisco and maybe she will be restored to airworthy condition in the VIP transport configuration from her time with the Hughes Tool Company. However, the condition of the airframe is reported to be marginal and has typical B-25 corrosion issues.

April 2005, she is now partially restored and has new paint. (John Shupek)

MILITARY AVIATION MUSEUM, VIRGINIA BEACH, VIRGINIA

The Military Aviation Museum is home to one of America's largest private aviation collections and includes marvellous flying machines ranging from WWI to the Vietnam conflict. It is located in Virginia Beach, Virginia. Each airplane has been beautifully restored to its prior military condition, using original parts whenever possible. Most of the airplanes are airworthy and flown at the museum during flight demonstrations and at air shows throughout the year. Some being the last flight-ready aircraft of their time, and all of them being one of a kind. The collection also includes a large reference library, along with artefacts and materials to illustrate the historic context of the aircraft in the collection. For nearly 20 years,

N7947C was stored for many years at the display yard of Walter A. Soplata at Newbury, Ohio. Below the cockpit is written "Our Miss Lady". (SmugMug)

Gerald Yagen has been assembling one of the world's largest collections of airworthy warbirds owned by a single individual. In 1969, Yagen opened a recruiting firm that matched military veterans with companies needing personnel with technical skills. That led him to take flying lessons. He felt flying was a convenient and more efficient way to travel on business. The heart of the collection was formed and created by Gerald and Elaine Yagen, long-time residents of Virginia Beach and founder of Tidewater Tech, now Centura College, and the Aviation Institute of Maintenance schools. The museum, which opened its doors in 2008, is situated on a private airfield with a grass runway and not less than eight hangars and shelters. It was created by Gerald Yagen, who's goal it was to show the younger generations these aircraft in their habitat, the sky, and perpetuate the remembrance of all those who had written combat history, sometimes giving their life to succeed.

One of the airplanes of the Military Aviation Museum is an incredibly restored and

in Newbury, Ohio. She was disassembled and moved to his display yard where she was stored for over 25 years. In December 1990, she was sold to Steven A. Detch of Alpharetta, Georgia. Finally, Gerald Yagen bought the airplane in December 1997. He restored the ship and her first flight after restoration was on 19 November, 2005. She flew as "Wild Cargo" but with a nice nose art.

A picture taken at Virginia Beach in January 2008 during maintenance. (WIX)

"Wild Cargo" of the Military Aviation Museum in full glory with a very nice nose art. Still flying at the time. (Collection Wim Nijenhuis)

flying B-25J, serial number 44-30129. This B-25J-25 was delivered in December 1944 and used as trainer and for administrative duties. In December 1957, she was flown to storage and in June 1958, she was sold to P. J. Murray of Oxnard, California and registered as N7947C. From September 1958, she was sold several times to different owners until January 1963, when she was sold to Arthur Jones of Slidell, Louisiana. She was named "Wild Cargo". The next month, when loaded with reptiles of a wild animal show, she made an emergency landing. The airplane was eventually auctioned off. In September 1964, Walter A. Soplata purchased the plane and took her to his house

Nowadays, she is displayed in the Army Hangar of the Military Aviation Museum.

MOVIELAND OF THE AIR MUSEUM, SANTA ANA, CALIFORNIA

Movieland of the Air Museum was a destination for movie buffs and airplane enthusiasts in the 1960s and 1970s. Two leading motion picture pilots of the era, Frank

Tallman and Paul Mantz, joined forces to service the film industry as well as showcase vintage aircraft at the Orange County Airport. See the chapter Tallmantz. The partnership of Mantz and Tallman formed the backbone of their business in providing Hollywood with airplanes for films, and also was the basis of the Movieland of the Air Museum which they opened at Orange County Airport on 14 December, 1963. The museum had a combined collection of airplane and movie memorabilia on display to the public. The first visitors found over

Exterior view of the new Movieland of the Air Museum taken in 1964 from across Campus Road looking west and an original sign of the museum. (William T. Larkins, Trevor McTavish)

fifty rare airplanes spread across a five-acre complex, both inside a large hangar and outdoors in a crowded display area. Surrounding rare World War I and World War II airplanes were a large aviation armament collection, uniforms, logbooks, movie posters, medals, trophies, and valuable mementoes. The museum was a popular attraction for several years. But after Paul Mantz' death in 1965 while filming the original "Flight of the Phoenix" movie, Frank Tallman continued the business, although some of the vintage aircraft collection had to be sold off. Tallman was forced to sell 45 airplanes and many display items to two Nebraska investors in January 1966. It was expected that the airplanes would shortly be resold and removed from the museum, and some were. But the bulk of the airplanes sold remained on display until May 1968, when the famed Tallmantz auction took place. After the auction, the airplanes sold left the museum, and Tallmantz closed the museum while new airplanes were being prepared for display. In 1969, the museum opened once again, though much of the museum area and the Tallmantz ramp were occupied by B-25s with paint schemes from the well-known movie "Catch-22" which were in storage and awaiting sale. Tallman died in an airplane crash in 1978 and the museum remained open during the remaining time of the company's operations under the original management. After Corporate President Frank Pine died in 1984, action was taken by the surviving Tallman and Pine families to liquidate the collection and sell the company. Most of the museum's collection was sold to Kermit Weeks in Florida, where much of it remains on display in his Fantasy of Flight museum. The museum was officially closed in 1985, and Tallmantz was sold to a new owner that same year. The company's new owner closed the doors definitively in 1991.

By 1969, much of the Movieland Museum area and the Tallmantz ramp at Santa Ana were occupied by B-25s with paint schemes from the movie "Catch-22". The airplanes were in storage and awaiting sale. (Kenneth Johnsen, Collection Wim Nijenhuis)

In addition to the Movieland of the Air Museum, an outdoor museum was established, called International Flight and Space Museum. Frank Tallman and Paul Mantz incorporated the International Flight and Space Museum in August 1962 as a non-profit museum to allow Tallmantz to obtain display airplanes from the U.S. government on loan. The museum was a fenced-in outdoor display area located near the Tallmantz hangars. The museum was used to obtain a significant number of military airplanes for display, including a B-50 fuselage and a

selection of missiles and rockets. By the late 1960s, the museum had largely been abandoned as a viable operation. The former display area was occupied by ex-Catch-22 B-25s being held in storage for sale, and many of the loaned military airplanes were hauled off to the south end of the airport for storage and were eventually reduced to derelicts. Within a few years most had been reclaimed by the government and sent on to other facilities, many, including the B-50 fuselage, going to the Planes of Fame Museum at Chino.

NATIONAL MUSEUM OF WW II AVIATION, COLORADO SPRINGS, COLORADO

A fine B-25 is operated by the National Museum of World War II Aviation. This museum houses a large collection of operational World War II aircraft and is located on a 20-acre campus on the northwest side of the Colorado Springs Airport. The National Museum of World War II Aviation was opened to the public on 27 October, 2012. The mu-

seum consists of three parts: a set of display cases with memorabilia, a connected hangar and work area and the adjacent hangar of WestPac Restorations, which has several airplanes that have either been fully restored or are in the process of being restored. WestPac Restorations is a full-service aircraft restoration and maintenance facility

and a premier restorer of WWII airplanes. Visitors to the museum may also take the few steps over to WestPac's restoration facilities to see flight-ready airplanes and airplanes in various stages of completion. This facility moved into to its current location on the north side of the museum campus in 2009 and after it was affiliated with the museum since its inception, it allows public access to its facility. Through close relationships with collectors around the country, the museum displays a wide array of fully restored aircraft that represent all aspects of America's involvement in World War II. Most of the aircraft have meticulously been restored to flying condition, and many are flown at the museum on public demonstration days. The museum is expanding and has plans for development of the Aviation Hall Project. This multi-phased expansion will feature a new aircraft display hangar and a building that will house new exhibit galleries, an events centre, and a state-of-the-art education facility.

N9117Z "In The Mood" is operated by the National Museum of World War II Aviation. She is overall natural aluminium finished and painted with the 345th "Air Apaches" Bomb Group markings. The beautiful nose art is only on the left side of the fuselage. (Collection Wim Nijenhuis)

The B-25 operated by the National Museum of World War II Aviation is N9117Z. This B-25J-20 was a former air tanker and sprayer in the 1960s. In the 1970s until 2012, she was sold several times and was named "In The Mood". On 18 April, 1992 she was launched off the USS Ranger for the 50th Doolittle Raid Anniversary together with the B-25 N8195H "Heavenly Body". On 29 August and 5 October, 1995, "In The Mood" was launched again from the carrier USS

Carl Vinson for the 50th commemoration of the end of World War II. In September 2000, she flew off the carrier USS Lexington and most likely also from the USS Constellation, for the movie "Pearl Harbor". In 2012, she was sold to Bill Klaers at Colorado Springs. Bill Klaers is co-chairman of the museum and is co-owner and President of Westpac Restorations. The bomber is still airworthy and is operated by the National Museum of World War II Aviation and is still flown as "In The Mood".

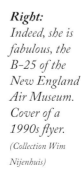

A photo of N3970C taken at Bradley International Airport in October 1975. At the time, she had a yellow tail number 34999. (R.A.Scholefield)

In 1992, "In The Mood" was launched from the USS Ranger for the 50th commemoration of the Doolittle Raid. In 1995, a sticker was applied next to the nose art in honour of the two launches from the carrier USS Carl Vinson that year. (NARA)

Right:
Indeed, she is fabulous, the B-25 of the New England Air Museum. Cover of a 1990s flyer.
(Collection Wim Nijenhuis)

NEW ENGLAND AIR MUSEUM, WINDSOR LOCKS, CONNECTICUT

She was restored by Tom Reilly and in 1986, the B-25 returned to static display at the New England Air museum. She is beautiful painted like her counterpart in World War Two of the 12th Bomb Group, 82nd Bomb Squadron. Now, she has a black tail number 34381.
(Collection Wim Nijenhuis, Eric Boisclair)

A beautiful B-25 is owned and exposed by the New England Air Museum at Windsor Locks, Connecticut. The New England Air Museum is owned and operated by the Connecticut Aeronautical Historical Association. In 1959, the Connecticut Aviation Historical Association was established to study and preserve Connecticut's aviation legacy. Eventually it amassed a large outdoor collection of aircraft across from the airport's main entrance on Route 75. In 1976, additional aircraft and displays were installed in one of the remaining World War II hangars. This became known as the Bradley Air Museum. Both sites were heavily damaged by a tornado in 1979. With support from the state, the museum and its salvaged aircraft moved to a 56-acre site on the west side of the airport and re-opened in 1981 as the New England Air Museum (NEAM). Today NEAM's collection includes more than 100 aircraft and thousands of artefacts in three larger hangars and an outdoor display. Located at Windsor Locks near Bradley International Airport, the museum is the largest aviation museum in New England. The Restoration Hangar is only open to the public on select days during the year. The Museum is governed by a volunteer Board of Directors and run by full-time employees, part-time employees, and volunteers. In 2017 work was begun on major improvements at the museum. A $1.9 million project aims to

enhance the visitor experience. The project will create lofty mezzanines in two of the massive aircraft hangars, which will provide vistas over the museum's aircraft collections. The renovations were unveiled in September 2017.

Within the aircraft collection of NEAM is the B-25H-10 with s/n 43-4999. After the war, she was transferred to the Dominican Air Force as FAD 2502 in 1950. In 1952, she was returned to the United States and her registration was assigned as N3970C. There was an attempted sale in November 1957 that was never completed, and the aircraft was derelict as of 1957. She was seized by

the Mercer County Airport for non-payment of fees. In 1970, she was donated to the Bradley Air Museum. At that point, she was disassembled and trucked to Bradley. There, she was restored to static condition for the first time. In October 1979, she was severely damaged in a tornado. She was restored by Tom Reilly in exchange for a B-17G airframe that was also damaged in the tornado. In 1986, the B-25 would be returned to static display at the New England Air museum. As of 2019, she is currently undergoing further restoration to return to static display including the reinstallation of her 75mm cannon.

PALM SPRINGS AIR MUSEUM, PALM SPRINGS, CALIFORNIA

A fine example of a warbird with colours of a B-25 that represents a ship of the 17th Tactical Reconnaissance Squadron, 71st Reconnaissance Group can be found in Palm Springs. The Palm Springs Air Museum is a first-class facility featuring a host of well restored and maintained aircraft including the B-25 "Mitch the Witch". The museum is a non-profit educational institution located on the north-east side of the Palm Springs

International Airport. The Museum's mission is to exhibit, educate and eternalize the role of the World War II combat aircraft and the role the pilots and American citizens had in winning the war. In addition to flying aircraft, related artefacts, artwork, and library sources are used to perpetuate American history. It contains a large collection of flying World War II warplanes and many of these airplanes have been used by

"Mitch the Witch" of the Palm Springs Air Museum around 2011, with the original paint scheme representing the B-25D which flew in the South West Pacific Area. (joolsgriff)

entrance to the museum is located in an attractive semi-circular glass wall. On either side of the entrance are mosaics or renderings that appear to be in stone. The mosaics depict American combat aircraft in flight. These appear along the wall and near the top of the two hangar structures that house the aircraft. To the left is the Pond Hangar Pacific Theatre and to the right is the Cravens Hangar European Theatre. Connected to the Cravens Hangar is the Phillips Hangar which houses the Boeing B-17G, the main-

motion picture companies in movies. The museum is housed in a new structure that includes three main display hangars, theatre, gift shop, ramp and airport access for flight demonstrations and visiting planes, research library, simulator, and education centre. In late 1993, the idea for a Palm Springs Air Museum sprang from the imaginations of three men: Charlie Mayer, Pete Madison, and Dr. Mort Gubin. The three men wanted to create an air museum with WWII warbirds in Palm Springs. Pete Madison, a former P-38 pilot, had just bought Robert Pond's home, and knew Pond had a collection of warbirds and classic cars in Minnesota. Pond was an industrialist, an aviator, a Navy pilot, an aircraft designer, a genuine car enthusiast, a philanthropist, a Minnesotan at heart and a man who turned a family business from eight employees to a global $100-million concern. He turned the company into one of the world's largest manufacturers of floor-cleaning equipment. He began his legendary collection of warbirds and other military aircraft in 1970. His B-25 Mitchell is one of the most flown

The ship in 2013. She has received another garish nose art that replaced the accurate former artwork she had. Nevertheless, she remains a very nice warbird. (Adam Duffield)

airplanes in the collection and appeared in the films "Forever Young" and "Pearl Harbor." Pond declared his willingness to help and with his support and contacts in aviation, the concept moved forward. The Palm Springs City Council approved and in January 1994, the Organizational Certificate of Incorporation of Palm Springs Air Museum, Inc. was signed. The construction started in February 1996 and the museum opened to the public on 11 November, 1996 with two hangars. A third hangar was opened in 1999, and a fourth hangar opened May, 2017. The

tenance shop, the Children's Education Centre, and the café. Located on the grounds on either side of the entrances are the Wall of Honour and the Distinguished Flying Cross Wall of Honour.

The B-25J-30 with s/n 44-86747 is nowadays known as "Mitch the Witch II". After USAF storage, she was sold to Alton C. Mosley at Fairbanks, Alaska and flew as a fire fighter with registration number N8163H. In December 1978, the bomber was sold to the Planes of Fame Air Museum in Chino and

Right side of the airplane with the insignia of the 17th Tactical Reconnaissance Squadron on her nose. The colourful vertical tails are painted as they were at the time of the war. (Palm Springs Air Museum, Collection Wim Nijenhuis)

in March 1986 to Robert Pond of Spring Park, Minnesota. He restored the airplane as "Mitch the Witch" and she flew with his collection for some years before being transferred and retired with the rest of the collection to the Palm Springs Air Museum, where she rests today. "Mitch the Witch's" paint scheme represents that of B-25D #42-87293, which flew in the South West Pacific Area with the 17th Tactical Reconnaissance Squadron, 71st Reconnaissance Group. In more recent years the aircraft, like several of the Palm Springs collection, has received inaccurate, garish nose art to replace the accurate artwork that she had. In 1986, she featured in the film "Forever Young". In September 2000, she was launched off the CV-16 USS Lexington and the CV-64 USS Constellation for the filming of the movie "Pearl Harbor". Currently, she is a static display at the Palm Springs Air Museum.

N2854G at Paul Bunyan Land in June 1992. The serial number 327102 on her tail is fake and her real serial number is 44–29812. She has the colours of the 310th BG on het vertical tails. (Jos Vervoort)

Here she is seen in August 1997. The airplane is named "Safe Return". (Collection Mike Laney)

PAUL BUNYAN LAND, BRAINERD, MINNESOTA

Apart from the aforementioned Big Kahuna's Water and Adventure Park, there was another park with a B-25 bomber. This was a B-25 surplus in 1958 and that was displayed for 42 years in Paul Bunyan Land, a tourist attraction in Brainerd, Minnesota. Paul Bunyan is a giant lumberjack in American folklore. His exploits revolve around the tall tales of his superhuman labours and he is customarily accompanied by Babe the Blue Ox. Paul Bunyan Land is a small family amusement park in Brainerd, featuring many rides for children and is surrounded by 30 buildings. The park, originally known as Paul Bunyan Amusement Centre, was founded in 1950 in Baxter, Minnesota, by

Sherm Levis. It was built around the statue of Paul that Sherm Levis and Roy Kuemicheal had purchased the previous year. The park grew over time to include over 40 rides. In 2003, the park announced that due to the high cost of running the park, it would be closing and that everything would be auctioned off. Fortunately, a local family-owned business named This Old Farm was interested in keeping the entire park and bought the statues of Paul Bunyan and Babe the Blue Ox as well the rides. They moved the park onto their land six miles east of Brainerd. In 2004, Paul Bunyan Land reopened after being relocated to this new location.

In July 2009, the airplane was transported to the National World War II Museum at New Orleans, Louisiana. She is displayed hanging from the ceiling. She carries the markings of the 490th Bombardment Squadron known as the "Burma Bridge Busters". (The National WWII Museum)

The B-25 airplane at Paul Bunyan Land was s/n 44-29812, a B-25J-20. After her USAF service as a trainer, she was stored at Davis Monthan in December 1957. She was sold to Paul Bunyan Amusement Centre in April 1958 and registered as N2854G. Probably this registration was only for the ferry flight to Brainerd. The bomber was displayed as USAF BD-812 and later as "Safe Return" in the colours of the 310th Bomb Group. For the visitors of the park, the airplane had a walkway on the right upper wing. In 2000, she was sold to Aero Trader and was dismantled and stored at Ocotillo Wells, California. In 2010, she was sold to the National World War II Museum at New Orleans, Louisiana. She was restored and equipped with a solid 8-gun nose and is displayed hanging from the ceiling in the U.S. Freedom Pavilion of the museum. The airplane is painted to represent a gunship of the 490th Bomb Squadron, 341st Bomb Group operating in Burma. The National World War II Museum focuses on the contribution made by the United States to Allied victory in World War II. The museum opened as the D-Day Museum, on 6 June, 2000, the 56th Anniversary of D-Day, focusing on the amphibious invasion of Normandy. It has been designated by the U.S. Congress as America's official WWII Museum. Powerful images and extraordinary artefacts bring to life the American Spirit, the courage, teamwork and sacrifice of the young men and women who won the war and changed the world. From the 1930s prelude to war, to the Normandy Invasion and the battles of the Pacific Islands, visitors trace America's role in the war and on the Home Front.

PLANES OF FAME AIR MUSEUM, CHINO, CALIFORNIA

The Planes of Fame Air Museum at Chino, California has always had a great collection of airplanes. The original collection consisted of just 10 airplanes. From those humble beginnings, the Museum has grown to the present number of over 150 aircraft. In the course of time, the museum also had four B-25s. These were N3339G, N3398G, N3675G and N8163H. The museum was founded by Edward T. Maloney in Claremont, California and opened on 12 January, 1957. At that time, it was simply called The Air Museum. Ed Maloney did not want to just display static aircraft in a museum, he also wanted to create a flying museum. Started with a flyable B-17, nowadays many of the museum's airplanes are also flyable. As the museum's collection of airplanes and memorabilia continued to grow, it became necessary to find a new home with enough space to house more airplanes. In 1963, the museum relocated to nearby Ontario Airport, Ontario, California. A few years later, in 1970, redevelopment of the airport at Ontario forced The Air Museum to move. The non-flyable aircraft became part of the "Movie World: Cars of the Stars and Planes of Fame Museum" in Buena Park, California, located near Knott's Berry Farm, while the flyable aircraft relocated to Chino Airport in Chino, California. When Movie World closed, the name Planes of Fame stayed. Planes of

N3339G at Ontario in October 1965. At the time, she was painted grey with a white fuselage top and a light grey fuselage bottom. (Nick Williams)

Fame consolidated in 1973, with its static aircraft joining the flying ones at historic Chino Airport. This was fitting, as the airport was formerly the home of Cal-Aero Academy, which was an Army Air Corps flight training facility. In 1995, an additional display facility was opened in Valle, Arizona. In 2004-2008, the Chino facility underwent further expansion with the construction of two new hangars, new offices, a gift shop, library, and a youth education centre and in October 2009, another new hangar was dedicated. The current President of the museum is Steve Hinton. He has been performing at air shows around the world for more than 35 years, flying over 150 types of aircraft and has restored more than 40 warbirds to pristine flying condition. Hinton has been President of Planes of Fame Air Museum since 1994. More about Steve Hin-

Here, N3339G is seen in Olive Drab/Neutral Grey colours at the Planes of Fame Air Show in Chino in 1978. (Chris Kennedy)

ton and his movie B-25 is described in the chapter American film industry.

N3339G

This was a special airplane. The B-25J-1 with s/n 43-4030 was delivered in February 1944. She was flown to North American Aviation in Inglewood, California to undergo an ex-

tensive modification to a VIP transport. The turret, nose gun, waist guns and tail guns were removed. Front to back, the bomb bay was lowered to accommodate a sleeping bunk. This required significant modifications to the wing carry-through. The remaining portion of the bomb bay was fitted with a fuel tank to extend the range. Aft of the bomb bay, seats, a full width table, and a walnut cabinet were installed. Windows were cut in the sides and the aft hatch was revised for easier access. The interior was insulated and covered in blue fabric over panelling. Commercial airliner seats were obtained from Douglas Aircraft. The former tail gun position held the relocated life raft. The new configuration could comfortably fit 10 people including the crew. After completion, the B-25 was ferried to the 8th Air Force in England. There, she was flown as General Dwight D. Eisenhower's personal transport. After the war, she returned to the U.S.A. and continued service until December 1958 when she was placed

Music video

In the days before there were real music videos, Charlatan Productions pioneered the form. Here is the Canadian American rock band Steppenwolf plus one go-go dancer standing atop a B-25 at The Air Museum in the late 1960s. Most likely, this bomber

is N3398G. Steppenwolf is best known for their hit "Born to Be Wild" in the opening scene of the film "Easy Rider" with Peter Fonda and Dennis Hopper riding their Harley choppers through America in the late 1960s.

(George Rocheleau)

Currently, s/n 43-4030 is displayed at Ellsworth AFB, South Dakota. The picture was taken in June 2006 at Ellsworth. Note the post-war engine modification of the 1950s. (Rob v. Ringelesteijn)

S/n 44-30761 was used as an instructional airframe at Shepard AFB, Texas in the early 1960s. Later, she got a black paint scheme. Although she looked quite complete and intact when photographed at Ontario in October 1970, the black N3398G was somehow deteriorated to an incomplete and in very rough condition. (Collection Wim Nijenhuis, R.A. Scholefield)

After "Shangri-La Lil", the B-25 was named "Betty Grable". (Collection Wim Nijenhuis)

N3675G here at Chino in September 1983 in Olive Drab and Neutral Grey colours and named "Shangri-La Lil". (Eduard Marmet)

The first nose art on "Photo Fanny" was painted by Teresa Stokes. In 2012, her nose art was revised. (Collection Wim Nijenhuis)

into storage. In April 1959, she was sold to Lanward Leasing Corp. of El Paso, Texas. Her registration was assigned as N3339G. After she had some different owners in the early 1960s, she was sold to Newell Hayes of El Paso, Texas in July 1965. She was donated to The Air Museum of Ontario, who transferred her to the USAF Museum Program in October 1981. She currently resides at the South Dakota Air and Space Museum at Ellsworth Air Force Base, South Dakota.

N3398G

This was a B-25J-25 with s/n 44-30761 and civil registered as N3398G. She went to Ed Maloney of The Air Museum in April 1964. She was donated by the USAF and was a former instructional airframe. She was stored in a weathered black scheme and marked as #30761. She was reportedly derelict and vandalized by the mid-1970s and subsequently scrapped. In 1987, the remains went to Carl Scholl of Aero Trader

Chino, May 2005, the bomber is still in flyable condition, but now in overall natural aluminium finish and named "Photo Fanny". The airplane is frequently used for film and television projects, both as a camera platform and as a subject for various photo projects. (Collection Wim Nijenhuis)

at Chino. The incomplete hulk stayed at the Aero Trader storage facility at Ocotillo Wells until 2000. From there, she went to the Central California Historical Military Museum at Firebaugh-Eagle Field, California. In April 2019, she went to Mitchell Aircraft Components Inc. at Borrego Springs, California.

N3675G

Carrying the serial number 44-30423, the airplane was built in the Kansas City plant of North American in 1944. She served with the USAF until the late 1950s when she was put up for disposal at Davis-Monthan AFB, Arizona. Purchased by National Metals in Phoenix, Arizona in July 1959, she was civil registered as N3675G. The B-25 was finally obtained by The Air Museum of Ed Maloney in the mid-1960s. The airplane has been kept in flying condition since the early 1960s. She regularly appeared at air shows throughout the Southwest and was frequently used for film and television projects, both as a camera platform and as a subject for various photo projects. So, the airplane has been both in front of and behind the cameras in many projects, films and TV-work. Over the years the airplane has worn different colours and names. From 1978, she flew as "Shangri-La Lil" and later as "Betty Grable". The last name "Photo Fanny" was adopted during the filming of "Iron Eagle III". The nose art was painted by Teresa Stokes and is based on nose art that carried originally by a photo-recon B-24 Liberator of the 2nd Photographic Charting Squadron in the Pacific Theatre of Operations. The artwork was inspired by art created by

Alberto Vargas. The ship appears in several films and flew in 2000 for the film "Pearl Harbor" as a camera ship and in the Hawaiian sequences as a Doolittle Raider with serial number 02261 in WWII USAAF camouflage. In Pearl Harbor she flew off the carrier USS Constellation. For filming Pearl Harbor a gyro-stabilized 35mm camera was put in the tail. She is the only fixed-wing plane ever to carry a gyro camera.

N8163H

The B-25J-30, s/n 44-86747, was flown to

N8163H was purchased by the Planes of Fame Air Museum in December 1978 which sold her after nearly eight years in March 1986. Here she is seen at Chino in the early 1980s.
(Larry Johnson, Collection Wim Nijenhuis)

storage in December 1958. In May 1959, she was sold to Alton C. Mosley from Fairbanks, Alaska and registered as N8163H. She was equipped with a fire-retardant tank and had several Alaskan owners. Her last Alaskan owner was Noel Wien of Anchorage, and he sold the bomber to the Planes of Fame Air Museum in Chino in December 1978. In 1986, she was featured in the film "Forever Young". The Planes of Fame Air Museum sold the B-25 in March 1986 to Robert Pond of Spring Park, Minnesota. Currently, she is on display at the Palm Springs Air Museum, California and flown as "Mitch the Witch II".

The dark green camouflaged "Buck U" of the Polar Aviation Museum in flight. (James Church)

POLAR AVIATION MUSEUM, ANOKA COUNTY, MINNEAPOLIS

Successful businessman and aviation enthusiast Wally Fisk decided to invest more of his time and capital into airplanes. Wally Fisk was an excellent mechanic, had a degree in engineering, owned a successful engineering and manufacturing company, and had already owned and flown civil airplanes for about 20 years as a private pilot. It was a time when many WWII aircraft owners who had just entered the scene were quickly discovering that the initial cost of the aircraft which they could afford, was just a down payment with the true operating cost being several times more. Therefore, operating old aircraft had to be run in a professional manner. With that in mind Fisk formed the AmJet Aircraft Corporation as an aircraft dealer and restoration shop. A few part time employees grew to two full time mechanics as his aircraft collection

area. The new museum facility was always open to the public, and especially returning veterans and families, just like the workshops had been in the past. But most of the time the huge building served as an indoor climate-controlled storage area for their show aircraft. In just a few short years the original two-man garage restoration shop had turned into a collection of 45 airplanes that were housed in 13 hangars, a full-time staff of around 15 people, a summertime air show schedule running 7 days a week, a wintertime restoration shop and a professional full-time operation that was run just like a large corporation. One of the significant airplanes in the collection was a B-25 which was flown as "Buck U". But despite all this, the Polar Aviation Museum was closed soon in 1999 and the organisation had to sell their collection which ranged from fly-

able planes in excellent condition to some restoration projects. The facility is now the Golden Wings Museum with some of the airplanes still in place.

The B-25 of Polar Aviation Museum was N3155G. This airplane flew with Max Biegert and Ontario Flight Service in the 1950s and 1960s. In 1976, she was purchased by Aero Trader and restored. She changed ownership again in 1980, in 1987, in 1990 and in 1992. Finally, Wally Fisk of Amjet Aircraft Corporation purchased the airplane in July 1993. She was operated at the Polar Air Museum in a flying status and named "Buck U". In August 1995, she flew off the aircraft carrier U.S.S. Carl Vinson to commemorate the 50th Anniversary of the end of World War Two. After the museum was closed, the aircraft was sold to Historic Aeroplane Works/Warbird Digest at Huntington, Indiana in February 2005. She was repainted as "Green Dragon". In December 2006, she was purchased by Claire Aviation and is currently flown as "Take-off Time" at Philadelphia, Pennsylvania.

"Buck U" in the hangar of the museum. Unfortunately, the museum closed its doors again a few years after its opening and the bomber was sold to Historic Aeroplane Works/Warbird Digest. (Ingo Warnecke via airport-data.com)

After "Buck U" was sold to Historic Aeroplane Works/Warbird Digest in 2005, she was repainted as "Green Dragon". One year later, the green camouflage paint was removed, and the bomber was natural aluminium finished. (Mark Silvestri)

expanded. Then it really began to expand quickly with the addition of more warbirds and professional specialists in the areas of maintenance, flight training, and overall administration. A next step was to bring the aircraft to the public and led to the formation of a subsidiary operation known as the Polar Aviation Museum. This museum was opened in 1996 and was based just across the field at the Anoka County Airport in Minneapolis. It had a large hangar with a main wide door leading out to the ramp

SST AVIATION MUSEUM, KISSIMMEE, FLORIDA

The SST Aviation Museum at Kissimmee, Florida owned two B-25s, N543VT and N9079Z. This museum was opened in 1973. SST was the first American Supersonic Transport (SST) project of Boeing. The SST was going to be America's answer to the British/

French Concorde supersonic jetliner. But under pressure from environmentalists who thought the SST's sonic booms and pollution would harm the atmosphere and those who thought the SST would be a waste of money, Congress killed funding for the project in early 1971. Boeing's full-size mock-up was disassembled and shipped from Seattle to Florida, where it was displayed at the SST Aviation Exhibit Centre literally built around it in 1973. Then on 4 July, 1973, the SST Aviation Museum opened to the public. Initially it drew big crowds. It had a nice collection of historical artefacts ranging from a Mercury space capsule to several rare World War II airplanes. But attendance steadily declined, and the museum closed in 1981. The aerospace and aviation artefacts on display eventually found new homes, except the SST. The plane was too big and expensive to move without destroying it or tearing down the building. In 1983, the building, complete with SST, was purchased by the Faith World church.

S/n 44–86698 was loaned to the SST Aviation Museum and at that time she was registered as N543VT. (aircraft–in–focus.com)

A colourful yellow, white, and red N9079Z outside the SST Aviation Museum in 1977. (John Hevesi)

N543VT

The B-25 bomber was a former Canadian airplane. Hicks & Lawrence Limited had owned this ex-RCAF B-25J-30, s/n 44-86698. She was registered as CF-NWU. After she had two other Canadian owners, she came back to the U.S.A. in 1967 and was registered as N543VT. She got several Texan owners. Daniel Jackson at Seymour loaned her to the SST Aviation Museum in 1978. After the museum closed, the Mitchell went to Canada again and was sold to G&M Aircraft at Alberta in August 1982. There, she was registered as C-GUNO and flew as a fire bomber.

N9079Z

After the B-25J-25, s/n 44-30734, was sold by Clements and Howe Aviation, Indiantown, Florida in January 1968, she was sold to Robert and Richard Howe in March 1973. At that time, the airplane operated as an insect sprayer called "Big Bertha". She was do-

nated to the SST Aviation Museum in 1974. She was flown to the museum and landed on a strip of highway blocked off by police. While taxiing to the museum, one engine of the old bird seized. She had to be towed the last few feet to her new home. After the museum closed, they sold her in October 1983 to Pat O'Neil, Robert Bolin, and Jack Myer of Wichita Falls, Texas. Currently, she is still airworthy as warbird "Panchito".

Flyer from the SST Aviation Museum.

(Collection Wim Nijenhuis)

Strategic Air Command & Aerospace Museum, Ashland, Nebraska

A unique display in the Strategic Air Command & Aerospace Museum, N3441G without wings and parts of the fuselage have been cut away to reveal the internal structure and interior. (AzAP)

Nose section of N3441G. (John Meneely)

A rather unique B-25 is displayed in the Strategic Air Command & Aerospace Museum at Ashland, Nebraska. The museum has two B-25s. One is s/n 44-30363 and is displayed in a clean Olive Drab/Neutral Grey USAAF camouflage scheme. But this airplane never saw civil service. The other B-25 is s/n 44-28738, and the airplane was civil registered. She is displayed without wings and parts of the fuselage have been cut away to reveal the internal structure and interior. It is, therefore a special and unique example to see. After this ship was stored at Davis Monthan AFB, she was sold to Whited Flying Service at Papillion, Nebraska in May 1959. She was registered as N3441G. In January 1963, she was sold to United Aerial Applicators Inc. at Papillion. She was to be converted to a sprayer but was not completed. She was stored at South Omaha, Nebraska until 1973. She changed ownership a few times and in 1975, she was transferred the USAF Museum at Offutt AFB, Nebraska. She was displayed as sectioned fuselage only. In May

1998, she went to the Strategic Air & Space Museum at Ashland and only the fuselage was displayed. The Strategic Air Command (SAC) & Aerospace Museum utilizes over 300,000 square feet of exhibit, education, and event space to educate. Established in 1959, the original mission of the museum commemorated the contributions of the Strategic Air Command, which was headquartered at Offutt Air Force Base until

1992. The Strategic Air Command Museum was located near Offutt Air Force Base in Bellevue. The airplanes on hand were located on an outdoor ramp. The airplanes were beat up; weathered looking and rusting from the weather conditions. Upon completion of a new indoor facility, the museum moved to its current location at Ashland in 1998 and today houses an impressive collection of military aircraft and space artefacts, many of which have been restored in the museum's on-site restoration facility. A wide range of traveling exhibits, permanent exhibits and special events provide unique experiences for museum guests. The name SAC Museum gave way to the Strategic Air and Space Museum in 2002, as the museum expanded its mission. In June 2015, the museum announced another name change to the Strategic Air Command & Aerospace Museum. The SAC Aerospace Museum is a registered non-profit organisation.

Texas Flying Legends Museum, Houston, Texas

The Texas Flying Legends Museum, with its fleet of twelve historic and award-winning aircraft, travelled across the country during the air show season bringing a unique viewing experience to the public. Hank Reichert started the museum with the purchase of

a P-51D Mustang in August 2010. A month later another P-51D joined the collection, also purchased by a private donor. Over the next five years, ten more aircraft would join the two Mustangs in the fleet. This fleet was based in hangars in Houston, Texas and

Minot, North Dakota. Each base was home to the aircraft over the course of the year. As the season's change the planes moved locations in accordance with the museum's appearance schedule. Unlike most other museums, what makes Texas Flying Legends so special was that the planes came to the people, rather than the reverse. The Texas Flying Legends collection had two owners/sponsors, Ed Bosarge and Bruce Eames. In 1998, both men were co-founders of The Quantlab Group. This is one of the world's most successful high-frequency trading organisations. Quantlab Financial, LLC develops and deploys (through affiliates) proprietary trading models used on dozens of exchanges and other trading venues around the globe. Unfortunately, Ed Bosarge decided to drop out of warbirds to pursue other interests. So, the Texas Fly-

ing Legends Museum disbanded at the end of 2017. The hangar at Ellington Airport in Houston was sold. The warbirds that Bosarge owned were sold and his B-25 "Betty's Dream" was the last to be sold. The rest of the former Texas Flying Legends Museum warbirds have been under separate ownership/sponsorship of Bruce Eames and have been and will be remaining together at the Dakota Territory Air Museum. The Dakota Territory Air Museum was always a sister museum to the Texas Flying Legends Museum, with the collection splitting time between both locations over the years.

The largest aircraft in the Texas Flying Legends fleet was the B-25J "Betty's Dream". This ex-Canadian airplane was registered

as CF-DKU and converted into a tanker. The airplane switched a couple of times in Canada and eventually ended up back in the U.S.A. with Aero Trader. Her civil registration was changed to N5672V in September 1994 and a restoration was started by Aero Trader. In March 1999, after restoration, she flew for the first time and was named "Betty's Dream". The airplane is painted in honour of Captain Charles E. 'Pop' Rice, Jr. of the 499th BS, 345th BG, assigned to "Betty's Dream" in June 1945. Co-piloted by Victor Tatelman, the original "Betty's Dream" escorted two "Betty" bombers carrying the Japanese peace envoy to Ie Shima on 19 August, 1945 and the return mission from the conference in Manila with General MacArthur's staff. By the time these talks ended World War II, "Betty's Dream" carried 22 mission symbols, including two silhouettes symbolizing sunken ships. Since 2011, the warbird was operated by Fighter Hangar 1, known as the Texas Flying Legends Museum, Houston, Texas. In June 2020, the airplane was for sale at Courtesy Aircraft Sales at Rockford, Illinois, for $2,950,000.

N25NA arrived at DuPage Airport, West Chicago, Illinois in 1993 to be operated by Jack Rogers/Air Classics Aviation Museum. *(Glenn Chatfield)*

THE OKLAHOMA MUSEUM OF FLYING, BETHANY, OKLAHOMA

A particular B-25 has had the distinction of serving with three foreign air forces: the Royal Canadian Air Force, the Venezuelan Air Force and the Bolivian Air Force. Currently, this airplane is operated by The Oklahoma Museum of Flying. This museum is dedicated to the preservation and operation of its based aircraft. The small but fine collection of this museum consists of two P-51 Mustangs, an Albatros L-39 jet, a Fokker Eindecker and the B-25. The museum is located at Wiley Post Airport in Bethany, Oklahoma. Owner of the museum is Dr. Brent Hisey. He is also pilot of the famous Reno Racing P-51 "Miss America" that he has been flying out of the Oklahoma Museum of Flying since 1993. Hisey is a neurosurgeon from Oklahoma City and as a child, loved building

plastic models of his favourite airplane, the P-51 Mustang. He always dreamed of flying one and now, he is getting his wish.

The B-25J-30, s/n 44-86725, was acquired by warbird enthusiast Brent Hisey in 2013. The bomber was named "Super Rabbit" and arrived at its new home at Wiley Post Airport in Bethany in 2013. She was transported overland on four tractor trailers. The B-25 was delivered in June 1945 and stored by the USAF. She was taken out of storage in October 1951 and transferred to the RCAF in January 1952 and was used for training and testing. She was removed from the RCAF inventory in October 1963 and sold to Columbus L. Woods of Lewistown Montana and registered as N92889. The same year, she

was sold to the Venezuelan Air Force along with eight other B-25s. She served with the Venezuelan Air Force for six years and was retired from active duty by 1972. In 1975, she was acquired by the Bolivian Air Force and served with the Bolivian Air Force for several years as an administrative and transport airplane and she made her last flight in l979. In July 1984, she was sold to C. A. Bird at Jacksonville, Florida and in May 1985, she was sold to Doan Helicopters of Daytona Beach, Florida. Her current registration N25NA was assigned. She was sold again several times and eventually sold to Brent Hisey in 2013. She has been with the Oklahoma Museum of Flying since undergoing a complete restoration. Her first flight since restoration was mid-2014. She is overall natural aluminium finished with a very nice nose art and still named "Super Rabbit". She has the markings of the 42nd Bombardment Group "The Crusaders" and 390th Bombardment Squadron. During the war, this group operated in the South West Pacific Area.

"Super Rabbit" arrives at the Oklahoma Museum of Flying in January 2013. She was transported overland on four tractor trailers. *(OMOF/Rich Lindsey)*

The B-25 is highly polished and has the markings and nose art of a ship of the 42nd BG "The Crusaders", 390th BS that operated in the South West Pacific Area during the Second World War. *(Collection Wim Nijenhuis)*

Tri-State Warbird Museum, Batavia, Ohio

A B-25 warbird in the U.S. with a rather non-American colour scheme is the Mitchell of the Tri-State Warbird Museum at Batavia, Ohio. The Tri-State Warbird Museum was completed and opened to the public in 2004. It was originally constructed as a hangar and museum display space at the Clermont County Airport. The Tri-State Warbird Museum was formed in 2003 by David O'Maley, with a commitment to preserve the aircraft of World War II, to educate visitors on America's role in WWII, and to remember those who fought for freedom, and honouring those who made the ultimate sacrifice. A Grand Opening First Annual Taking Flight Fundraising Gala was held in May 2005. In 2011, an additional hangar and storage/shop space was built. This became the "Bomber Hangar" with space to display the aircraft and provide a wonderful educational tour. Continuing to grow, the Tri-State Warbird Museum aircraft collection consists of nine significant WWII aircraft with additional acquisitions planned. All the aircraft and other vehicles within the museum are in full operating condition.

The B-25J-35 with s/n 45-8898 was the second last B-25 bomber built. She is nowadays flying as "Axis Nightmare". The airplane was delivered when the North American Aviation B-25 contract was terminated. She was delivered to the RFC from the factory. In December 1946, she was added to the USAAF inventory and flown to storage. In 1948, after storage she was recalled for maintenance and eventual administrative and utility duties. In December 1958, she was stored at Davis Monthan. In August 1959, she was sold to H. H. Coffield of Rockdale, Texas and registered as N3681G. By August 1974, she was derelict and stored. In October 1983, she was sold at an auction to Aero Trader and trucked to Ocotillo Wells, California for storage. Binary Warriors of Weston Massachusetts purchased her in July 1985. She was disassembled and trucked to Tom Reilly, Kissimmee for restoration. Reportedly, her centre section was left at Aero Trader and the centre section of s/n 44-28765 was used for restoration. In November 1985, she received her current registration of N898BW. Her first flight after restoration was in 1988. She was finally sold in 1991 to Gene Rayburn of Arizona who operated her for several years in an unusual basic RAF colour scheme. The British RAF operated the B-25 in Europe during World War Two, but never in the three-tone camouflage scheme as carried by this restored airplane. In July 1997, she was sold to Don George of Springfield, Illinois. Her RAF scheme was altered to represent American markings and she was renamed "Axis Nightmare". In April 1999, she was transferred to Warbirds Incorporated at Springfield and in 2004, she was sold to the Tri-State Warbird Museum at Batavia, Ohio.

This is the last but one B-25J, serial number 45-8898, and also the last but one B-25 bomber built by North American Aviation at Kansas City. She went directly to the Reconstruction Finance Corporation for disposal and was being stored at Altus, Oklahoma. She is seen here in her post-war U.S. Air Force service of Headquarters Command at Bolling Field. She is current registered as N898BW. (USAF)

N898BW at the ramp of Tom Reilly in 1991. (Collection Wim Nijenhuis)

Right: *In a later stage, she picked up invasion stripes, U.S. insignia, and the name "Axis Nightmare". Magnificent photograph of "Axis Nightmare" of the Tri-State Warbird Museum in a flight around the National Museum of the U.S. Air Force at Wright-Patterson Air Force Base, Ohio, 18 April, 2010. (U.S. Air Force/Tech. Sgt. Jacob N. Bailey)*

VICTORY AIR MUSEUM, MUNDELEIN, ILLINOIS

The Victory Air Museum was formed by Earl Reinert and Paul Polidori. The museum was located near Mundelein, Illinois on Gilmer Road. Reinert was an early warbird collector. Lots of airplanes flying today were saved in whole or in part by Reinert. It was not a museum in the modern sense but more a collection of saved airplane stuff. This was a collection of about a dozen World War II fighter planes. Reinert and Polidori opened the private airstrip and air museum in 1961. At one time they had more than 20 World War II fighter airplanes. Among them were eight German Messerschmitt Me-109s, two British Spitfires and a Japanese Kamikaze, along with American airplanes as a Hellcat, Lodestar, B-25 and A-26. The Victory Air Museum no longer exists. It closed following the death of Paul Polidori in a flying accident in June 1985. One of the airplanes

In the 1970s and early 1980s, N17666 was displayed at the Victory Air Museum at Mundelein. She was named "Tokyo Express" and a shark nose was painted on her nose. By the end of the early 1980s, the airplane was in a deteriorated condition. (WIX, Collection Wim Nijenhuis)

Nowadays, N17666 is displayed at Pendleton Army Airfield Gate. (WIX)

WARBIRDS OF GLORY MUSEUM, BRIGHTON, MICHIGAN

of the museum's collection was the B-25 with the civil registration N17666. This was a B-25J-25 with s/n 44-30243. After USAF service this airplane was stored at Davis Monthan AFB. In June 1958, she was sold to Maricopa Dust & Spray at Maricopa, Arizona and registered as N9622C. Thereafter she was sold several times to various companies. In 1967, she was sold to R. McPherson/ Aero Dix, New Albany, Ohio and registered as N17666. She was withdrawn from use and placed in open storage at Elkhart, Indiana. In 1971 she was sold to Earl Reinert of the Victory Air Museum. She was delivered by truck and restored for static display as "Tokyo Express". After the closure of the museum, the B-25 went to Jay Wisler of Warbird

Parts & Memorabilia, Tampa, Florida and in 1994, to Tom Reilly Aviation at Kissimmee, Florida and Darryl Greenameyer, Ocala, Florida. She was dismantled and restored for static display. In 2000, the bomber went to the USMC Air Museum, MCAS Quantico, Virginia and was loaned to B-25 Preservation Group at Pampa, Texas. She was painted blue and marked as a PBJ-1. In 2002, she was moved to the Pendleton Air Museum, Pendleton, Oregon and is nowadays displayed at Pendleton Army Airfield Gate.

This museum was founded in April 2013, by Patrick Mihalek assisted by Todd Trainor, both with a passion for aviation. The primary objective is the preservation of historical aircraft and WWII memorabilia for the benefit of future generations. The goal of the museum is to restore the B-25 #44-30733 to 340th Bomb Group markings and fly her in air shows. The restoration of the B-25 is a multi-year project, requiring extensive work. An important mission of the Warbirds of Glory Museum is their youth mentoring programme and the Warbirds of Glory Museum is leading in this field. By 2018, the staff has mentored more than 30 kids, teaching them trade skills, shop discipline, toolmanship, and machining

The tanker N9088Z forty-four years after she landed on the sandbar in the middle of the Tanana River in Alaska, she was recovered in 2013.
(Warbirds of Glory Museum)

The well-restored glass nose of the Sandbar Mitchell.
(EAA/Sara Nisler)

Adventure Park at Destin. In 2005, she was removed from the pole and sold. Nowadays, her forward fuselage section is being used to rebuild the Sandbar Mitchell.

B-25 #44-28898 was recovered by the Nome Aviation and Military Museum. She included the centre section, forward fuselage, wings, partial rear fuselage, engine mounts, and cowling. This airplane never got a civil registration number. She was given to the USSR under the Lend-Lease Programme. In 1944, she suffered a serious landing accident when she arrived at Nome, and her repair was beyond the means of the rudimentary facilities at the base. So, the wreck found her way to the dump, where a wartime salvage crew removed useable parts, before abandoning the rest to time. Thereafter, she suffered further indignities from locals and visitors alike. Parts have wandered off, initials carved, and more than a few people have taken pot-shots at the remains. In early 2000's, the locally-based Nome Aviation and Military Museum decided to recover the wreck from its dump site. The Warbirds of Glory Museum took ownership of the remaining airframe in 2015. The spar caps have signs of corrosion, so she is not an airworthy airframe as is, but great candidate for a static restoration. They are using a few of the good extruded parts such as the fuselage attachment angles to restore the Sandbar Mitchell. In turn, the attachment angles that were damaged from the Sandbar Mitchell that are in static condition will go with #44-28898.

equipment operation, which includes CNC, computer-aided design engineering, and aviation maintenance. Most of the students that have been a part of the museum's programme follow a career in the aviation industry, mechanical skills industry, or engineering.

As mentioned before, the museum owns the B-25 with s/n 44-30733, registered as N9088Z. For the restoration they not only make new parts, but also use parts from two other B-25s. One is s/n 44-30947 and the other is s/n 44-28898. In 1969, airplane N9088Z made a wheels up landing on a sandbar in the middle of the Tanana River in Alaska. On 5 July, 2013, she was recovered by the Warbirds of Glory Museum and transported to the museum. Named "Sandbar Mitchell", the bomber is now undergoing restoration by the museum. The restoration is being accomplished by skilled volunteers, as well as the museum's youth mentorship programme. Each section of the aircraft is being carefully disassembled and repaired to the original factory drawings. Currently, the glass nose has been restored and progress is being made on the fuselage centre section.

The B-25 with s/n 44-30947 registered as N92880, was displayed on a pole above the water rides in Big Kahuna's Water and

Left: *The B-25 #44-28898 had been abandoned at Nome for more than seven decades.*
(Warbirds of Glory Museum)

In 2018, the Warbirds of Glory Museum displayed the forward fuselage section of the Lend-Lease bomber at the EAA AirVenture at Oshkosh. (Robert Domandl)

WILEY SANDERS WARBIRD COLLECTION, TROY, ALABAMA

In 1990, when owned by Wiley Sanders, she had a glass nose and was painted grey with "Ol Gray Mare" nose art. The letters W and S of Willy Sanders are on the vertical tail surfaces. (Bill Word)

Around 1967, N3438G had a solid nose and was overall natural aluminium finished with a white fuselage top. (Collection Wim Nijenhuis)

The Sanders warbird collection in Troy, Alabama has always been private although visitors are sometimes welcome in small numbers during business hours by a pre-arranged appointment. In the privately owned warbird collections, the Sanders family's warbirds were an actively flown collection for many years. Although the airplanes remain in pristine condition, they are not often seen publicly. The warbirds belonged to Wiley Sanders owner of Wiley Sanders Truck Lines. Wiley Sanders is a major hauler of plastics, industrial chemicals,

lime, and cement, with each tanker in their fleet being dedicated to a particular product and industry to eliminate possible products mixing. Wiley Sanders Truck Lines also do maintenance and repair. Wiley Sanders began his trucking company in 1959 with one truck. Since then, Wiley Sanders Truck Lines has grown to become one of the nation's leading common and contract motor carriers. His trucking company eventually grew to 600 trucks and 1,600 trailers and led to the growth of a business that operated in nearly every state of the U.S. Wiley Sanders

established the second of his companies in Troy in 1970, when he incorporated Sanders Lead Company. His purpose was to recycle the lead plates extracted from batteries and provide a product his trucks could haul back to Troy after completing freight deliveries. This business proved to be ultra-successful and led to many expansions. Operating in 48 states, Canada and Mexico, Sanders Truck Lines maintains mutually profitable relationships with hundreds of the nation's most demanding shippers. Wiley Sanders maintains a quite impressive collection of classic trucks and airplanes as well. The warbird collection of Sanders is located at Troy Municipal Airport. Since the 1980s, Sanders owns two B-25 Mitchells with the civil registrations N3438G and N5256V.

N3438G

The B-25J-30, s/n 44-86797, was flown to storage at Davis Monthan in October 1957.

N5262V at Oshkosh in 1983, just after the airplane was bought by Wiley Sanders. (Glenn Chatfield)

A few years later she had the Sanders S painted in a white triangle on her vertical stabilisers. The glass nose has been restored and the name "Georgia Mae" has been applied. (Glenn Chatfield)

George Race at Dallas, Texas purchased her in May 1959, and she was registered as N3438G. In the following years, she had several owners and by August 1983, she was sold to Aero Trader at Borrego Springs. In February 1985, after restoration, she was sold to Wiley Sanders Truck Lines. At first Sanders flew the B-25 with the name "Samantha", and later as "Ol Gray Mare". In June 2004, the bomber was sold to Hans Lauridsen of Glendale, Arizona and currently she resides at the Lauridsen Aviation Museum.

N5262V

The B-25J-30 with s/n 44-86785 is nowadays flying as "Georgia Mae". After her USAF service, she was stored at Davis Monthan by October 1957. In January 1958, she was sold to C. T. Jensen of Tonopah Air Services in Tonopah, Nevada. Her civil registration was assigned as N5262V. In March 1975, she was sold to Mid-Pacific International Inc. at Eugene Oregon and in October 1980, she was sold to Gary H. Flanders of Oakland, California. Finally, Wiley Sanders Truck Lines purchased the airplane in April 1983. Wiley Sanders restored the B-25 in the early 1980s and named her "Georgia Mae." She is still airworthy and privately owned by Sanders.

"Georgia Mae" in the hangar at Troy in November 2016. (David Ilott)

WINGS OF HISTORY MUSEUM, SAN MARTIN, CALIFORNIA

A number of B-25s in the world are still waiting for restoration. One of them is in the Wings of History Museum. This museum is an all-volunteer, non-profit organisation dedicated to the preservation and restoration of antique airplanes and displaying items from aviation history. The museum has been in existence for more than 25 years. The name has changed over the years, but the spirit and enthusiasm of the members has always been the driving force that makes the museum what it is today.

In October 1969, N3161G was sold to Archaeopteryx Corp. at Minneapolis, Minnesota where she flew with the Southern Minnesota Wing of the Confederate Air Force as "I See No Problem".

(Richard W. Kamm)

They curate exhibits of more than 50 models, engines, and airplanes, with a strong emphasis on pre-1970s technology and the evolution of flight. The museum also has a restoration shop, library, and propeller shop which specializes in repair and refurbishment of wood propellers.

The museum is located at San Martin, California, adjacent to the San Martin Airport formerly known as "South County" airport. The Wings of History Museum has two large hangars filled with airplanes and instrument parts. Hangar 1 contains mainly antique airplane parts, like engines, and a full-size, complete albeit non-flying replica of the Wright Brothers' original Wright Flyer. Hangar 2 contains mainly complete airplanes, and a motorcar-helicopter that was donated. Outside the hangars, there are some other airplanes including a B-25 Mitchell. This airplane is a B-25J-25 with s/n

Around 1992, she was renamed "Snow White II" and was owned by Dave Tallichet. (Glenn Chatfield, Collection Wim Nijenhuis)

44-30324. After storage at Davis Monthan, she was sold to Aviation Rental Service, St. Paul, Minnesota in October 1958. She was registered as N3161G and stored unconverted until 1965. In July 1965, she was sold to Milford O. Gillett at St. Paul and in November 1967, she was sold to Robert J. McManus of St. Paul. In October 1969, she was sold to Lynn L. Florey & Donald W. Ericson/ Archaeopteryx Corp. at Minneapolis, Minnesota where she flew as "I See No Problem". She was sold again in 1978 and in 1979. In 1981, she was sold to Dave Tallichet/Military Aircraft Restoration Corp. at Chino, Califor-

nia. By 1992, she was named "Snow White II". She was loaned to Empire State Aerosciences Museum at Schenectady County Airport, New York and in 2003 to MAPS Air Museum, Akron, Ohio. Finally, in 2007 she was ferried from Ohio to the Ken McBride Collection at San Martin. In June 2011, she was loaned to the Wings of History Museum at San Martin. She was stored waiting for restoration. In July 2020, the old solid nose was found in California by members of the B-25 History Project. The nose was transferred to Kansas City and will be restored by the B-25 History Project organisation.

In July 2020, the old solid nose was found and transferred to Kansas City and will be restored by the B–25 History Project. (Dan Desko)

The bomber is now equipped with a glass nose. November 2012, the B-25 between the hangars 1 and 2 of the Wings of History Museum is in a deteriorated state. The vertical stabilisers and rudders have been removed. (Rob Lockhart)

YANKS AIR MUSEUM, CHINO, CALIFORNIA

The Yanks Air Museum at Chino, California has one B-25 since 1999. This airplane with s/n 44-86791 was bought in Australia, then registered as VH-XXV. The museum shares Chino Airport with the Planes of Fame Museum. Yanks Air Museum houses one of the largest and most historically significant collections of American aircraft including World War II fighters, dive, and torpedo bombers. The museum was started by Charles and Judith Nichols in 1973, who were developing a lumber, property, and industrial business in town. They started with a single airplane, a Beech Staggerwing. They started the collection with the intent of trying to preserve this history for future generations. Today the collection contains over 200 aircraft with some being the last survivors of their type. There are two indoor hangars, one outdoor hangar and one restoration hangar, at the museum. The Chino facility encompasses 16,400 m² under roof and covers 40,000 m². In addition to the display hangars, public access is permitted, on a supervised basis, to the main restoration hangar and boneyard where historic aircraft are in various stages of restoration. The indoor hangars features most of its collections, many are the last surviving examples of their kind. Most of the aircraft in the museum are in airworthy condition. During air show and museum events, they are up in the air. The museum aircraft have been restored in-house with great attention to detail so that both flying and static aircraft are considered exemplary examples of their type. Great care and effort have been made to conform to the original manufacturer's specifications in all restoration projects. Another museum facility is being built in Greenfield, California, which will also include an advanced-technology education centre, hotel and spa, winery, restaurants, service facilities, shops, and a recreational vehicle park.

The B-25 of the Yanks Air Museum was purchased in March 1999 from the Australian War Memorial. She was shipped from Sydney to Chino to undergo an extensive restoration. In May 2002, she received her current registration N6116X. After the restoration, the bomber made her first flight on 22 June, 2002. The bomber is still adorned with a beautiful tiger. This painting was

from the logo of ESSO, which sponsored the delivery flight from the U.S.A. to Australia when the Australian War Memorial bought the airplane. Currently she is a static display in the Yanks Air Museum.

N6116X of the Yanks Air Museum at Chino. (David Biscove)

YANKEE AIR MUSEUM, YPSILANTI, MICHIGAN

The Yankee Air Museum was founded in 1981. A group of aircraft enthusiasts, adopting the name Yankee Air Force, shared the desire to preserve the facts and glamour of south-eastern Michigan's aviation history. They began to lay plans to research, restore, and preserve the all-but-forgotten history of Willow Run Airport, its role in the Second World War, and the historic aircraft of that and succeeding eras. Their initial goal was to acquire one of the original U.S. Army Air Forces hangars at the airport and restore it to original condition. The museum grew quickly when it acquired a flyable B-17 Flying Fortress, B-25 Mitchell, and C-47 Skytrain. Since 1981, the Yankee Air Museum

has acquired and returned to flying status six historic aircraft. The museum was originally located on the very North-East side of the Willow Run Airport, Ypsilanti until it suffered a devastating fire in October 2004. Fortunately, a few aircraft, including the B-25, were rescued. After the fire, the Yankee Air Museum began the process of rebuilding and relocating the museum, reopening in October 2010 at its current location in a repurposed building on the east side of Willow Run Airport. But this facility was never large enough to house all the museum's exhibits, collections, and its fleet of flyable vintage aircraft. In 2011, negotiations began for acquiring part of

the old Willow Run Bomber Plant that produced B-24 bombers during World War Two for the Ford Motor Company. After nearly two years of negotiations, an agreement was made to save a portion of the bomber plant and a financial campaign was raised to create a new, permanent home for the museum, its exhibits, educational programs as well as its flyable and static aircraft. When the Yankee Air Museum moves into a preserved and renovated portion of the former Willow Run Bomber Plant after the bomber plant has been restored and renovated, it will become the National Museum of Aviation and Technology at Historic Willow Run.

The Yankee Air Force B-25D is one of the very few existing early B-25 models. She is a B-25D-35 with s/n 43-3634 and was delivered in December 1943. On paper she was assigned to a Lend-Lease program for the

In the 1970s and 1980s, N3774 flew as "Gallant Warrior". She still had the solid nose which was installed by the RCAF as well as the white fuselage top. (Richard W. Kamm)

"Yankee Warrior" of the Yankee Air Museum at Grimes Field, Urbana in April 2017. She was restored to military configuration in 2000–2003 and fitted with a glass nose, replacing the solid nose. (Dennis Nijenhuis)

RAF as 'KL148', in January 1944. Instead, she was assigned to the 340th Bomb Group in Italy and flew eight combat missions with the tail number 9C between April and May 1944. The airplane returned to the United States in May 1944 and was delivered to the RCAF where she was used for training. Because of landing accident damage, her original greenhouse nose was damaged, and a solid nose was installed. In June 1962, the airplane was removed from the RCAF inventory and sold to Hicks and Lawrence from Ostrander, Ontario. A provisional registration was assigned as CF-NWV for the ferry flight. In June 1968, she was sold to Richard McPherson from New Albany, Ohio and in July 1968, she was sold to Glenn Lamont from Detroit, Michigan and registered as N3774. She was restored to airworthy condition and flew with a new FAA certificate as "Gallant Warrior". In October 1988, she was purchased by the Yankee Air Force. She was restored to military configuration in 2000-2003 and fitted with a glass nose, replacing the solid nose and flew as "Yankee Warrior". In December 2020, she received a new Olive Drab/Neutral Grey paint scheme and in March 2021, her new name and nose art was unveiled: "Rosie's Reply".

Detail of the fine nose art and the cockpit interior. (Wim Nijenhuis)

In March 2021, she received a new name and nose art. (Yankee Air Museum)

U.S. MILITARY BASES

Apart from the B-25s in the private museums, there are civilian B-25s that have ended up at various military bases in the U.S. Mostly Air Force bases, but also some other military places or former military places. They are displayed there in exhibition spaces or outdoors in the open field. Often, they are placed as a monument to commemorate the involvement or participation of the base or its units in the Second World War. Hereafter, a summary is given of civil B-25s that are currently being displayed at military locations or former military locations in the U.S.

CASTLE AIR MUSEUM, ATWATER, CALIFORNIA

Castle Air Museum is a military aviation museum located in Atwater, California adjacent to Castle Airport, a former Air Force base which was closed in 1995. Castle Air Museum has over seventy restored vintage military airplanes. The museum was officially opened to the public in 1981 as a non-profit organisation with a mission "to preserve military aviation heritage for future generations." Castle AFB had been a Strategic Air Command bomber base prior to its closure. The B-25 at Castle is a B-25J-30, s/n 44-86891. She has the civil registration N3337G. In the 1960s, she was converted into an air tanker. In January 1980, she was transferred to Castle AFB. During the 1980s, the B-25 was fitted with a greenhouse nose and was on display with Doolittle Raider markings. She has since been refurbished and repainted as a 345th Bomb Group airplane with the new name "Lazy Daisy Mae" with nose art of the Li'l Abner comic strip character.

After military service, N3337G was initially converted into a tanker for firefighting. She has been on display since 1980 at the Castle Air Museum at Atwater. She is seen here in March 2016. (Alan Wilson)

Grand Forks AFB, North Dakota

N9865C at Galveston, Texas in the 1980s. Pilot Jim. W. Hazlitt at Galveston was one of the owners of the B–25 in the 1970s. His name was painted below the cockpit window. (John Kerr)

At the front gate of Grand Forks Air Force Base, a few restored airplanes are on display including a renovated B-52, KC-135, Minuteman III, F-101 and a B-25 Mitchell. The name "Flo" emblazoned the side of this B-25 next to an image of a leggy dark-haired beauty, lounging luxuriously in a long white gown. This B-25 monument is not the same airplane as the original of Grand Forks that

In 1985, the bomber was donated to Grand Forks AFB and was displayed as "Flo", to honour the crew of a bomber with the same name of the 321st Bombardment Group during the war.
(Steven M. Dennis, Earl Leatherberry)

is N9865C, the B-25J-15 with s/n 44-28834. In May 1960, she was converted into a fire tanker and thereafter, until 1985, she had different owners. In 1985, she was taken up in the USAF National Museum Loan Program and loaned to Grand Forks AFB and was displayed as "Love Machine" and later as "Flo".

native John H. O'Keefe flew during World War II. But this bomber does bear her tail number and nose art and is meant to honour the crew of the bomber of 321st Bombardment Group, a group that later evolved into the 321st Missile Wing stationed at the local base. Grand Forks AFB is located in the heart of the Red River Valley at the junction of the Red Lake River and the Red River of the North and plays a central role in the U.S. defence. It is home to the 319th Air Base Wing, the only base in Air Mobility Command to receive remotely piloted aircraft systems. The Mitchell of Grand Forks

Grissom Air Museum, Peru, Indiana

The Grissom Air Museum, near Peru, Indiana is named for astronaut Virgil I. "Gus" Grissom. The museum was founded in 1981 by seven prior service military personnel who lived in the area. The Heritage Museum Foundation (HMF) wanted to preserve aircraft that were currently located at Grissom Air Reserve Base, formerly Grissom Air Force Base. The HMF started the Grissom Air Museum in 1982 outside of what then was the main gate of Grissom Air Reserve Base. The

indoor museum was completed in 1991. The original airplanes were moved from the base to a public access site in 1987. Located at Peru, Indiana, the Grissom Air Museum has an impressive collection of military airplanes, including a B-25. There is a small museum building at the entrance packed full of uniforms, posters, bombs, and cockpits for everything from fighter jets to Huey helicopters. The highlights are the more than 25 airplanes outdoors. One of them

is a B-25 survivor of the movie "Catch-22". In 1959, this airplane was registered as N3507G. In the late 1960s, the airplane was purchased by Tallmantz and used in the film "Catch-22" where she flew as "Passionate Paulette". After filming, the airplane was donated to Grissom AFB in 1972. There she was also displayed as "Passionate Paulette". In the nearly five decades that she has been displayed at Grissom, she had several different paint schemes.

HILL AEROSPACE MUSEUM, OGDEN, UTAH

A clean machine is displayed at the modern museum of Hill Air Force Base. Hill Aerospace Museum is located on approximately 30 acres on the northwest corner of Hill AFB, about five miles south of Ogden. It was founded in 1981 as a part of the USAF Heritage Program and first opened in 1986. It moved to its current modern facility in 1991. Not only does the museum have close to 100 airplanes displayed in its two inside galleries and outside air park, it exhibits thousands of artefacts depicting the history of aviation of the USAF, Hill AFB, and the State of Utah. In 1996, the museum became the home of the Utah Aviation Hall of Fame. A B-25 bomber that is part of the museum's collection is a B-25J-30 with s/n 44-86772. The B-25 was accepted by the USAAF in June 1945 and was immediately placed into storage. She was moved around between various maintenance and storage fields in California, Missouri, and Texas. In 1958, she was transferred to Davis-Monthan AFB for long term storage. She was declared surplus to the USAF needs and was sold to National

Metals in Phoenix, Arizona in July 1959. She was registered as N9333Z. In November 1960, she was sold to Frank Froehling, Coral Gables, Florida and in October 1961, to David W. Brown at Hialeah, Florida. In January 1962, the airplane made a forced landing in a farmer's field in Argentina after suffering either engine problems or running out of fuel. The airplane was apparently being used to smuggle cigarettes into Argentina from Paraguay at the time of the incident. She landed in a rough field and the nose wheel collapsed causing damage to the front of the aircraft. She was then donated to a local flying club, where she was moved to Villa Canas Aero Club for display. Then a letter was sent to the FAA in November 1964 requesting that the airplane's registration be cancelled since she was perma-

In 2006, the airplane got new markings and was painted Desert Sand with a Neutral Grey underside. (Karl Hauffe)

nently out of service. The registration was cancelled the next month and she sat at the small airfield in Argentina for the next 27 years. In 1990, Don Whittington of Fort Lauderdale, Florida obtained the airplane, and she sat in pieces in his hangar until 1993. The airplane was reassembled and restored, and Whittington traded the B-25 to the USAF Museum System for four H-1 helicopters. The Mitchell was assigned to the Hill Aerospace Museum for permanent display and was painted to resemble the B-25s flown by the 345th Bombardment Group "Air Apaches".

N3507G is photographed in 2017. Again, her paint scheme has been changed. She is now painted Olive Drab and Neutral Grey, although the latter looks more like white. (GAM).

A clean machine, this Mitchell at the Hill Aerospace Museum. She had the civil registration N9333Z and the vertical tails are painted in the colours of the "Air Apaches".
(Bubbinski, Darth Dog)

HURLBURT FIELD MEMORIAL AIR PARK, OKALOOSA COUNTY, FLORIDA

Hurlburt Field is located on the Gulf of Mexico in the Florida Panhandle, 35 miles east of Pensacola. Hurlburt Field is close to Eglin AFB. This area is also known as the Emerald Coast and is a major tourist attraction for its breath-taking white beaches and the emerald green waters. Hurlburt Field is home of the Air Commandos since 1961, Hurlburt Field today accommodates the 1st Special Operations Wing, Headquarters Air Force Special Operations Command, an Air Force major command, and a number of associate units. The B-25 at Hurlburt is airplane #43-28222, a B-25J-10. This airplane was assigned to various units until being placed in storage at Davis-Monthan AFB in October 1957. She was purchased by Les Bowman in January 1958 and registered as N5256V. In the 1960s, she flew as a fire fighter and sprayer. In 1980, she was transferred to Beale AFB, California and displayed as #328222. In April 1995, she went to Hurlburt Field. She was restored for static display and displayed as #28222. This B-25 was reconfigured to resemble the famous B-25H-model of the 1st Air Commando Group in the CBI Theatre. The B-25 is part of a fantastic collection of military airplanes used by "Air Commando's" over the last century. This outdoor collection is displayed at Hurlburt Field Memorial Air Park. There are also many stone memorials to the fallen "Air Commandos" who lost their lives in combat. Hurlbert Field Memorial Air Park is located adjacent to Hurlbert Field and is publicly accessible after first checking with the base security at the gate.

MALMSTROM AFB MUSEUM, GREAT FALLS, MONTANA

The B-25 that played the leading role in the film "Catch-22" is currently displayed in the Malmstrom AFB Museum. This museum is located at Great Falls, Montana and focuses on the history of Malmstrom Air Force Base. Shown are a reconstructed barracks from the Second World War, weapons, uniforms, and other objects. Malmstrom Air Force Base was established in 1942. During World War II, B-25 bombers arrived at the base by rail and were assembled on the base, others were flown in by both military and Women Air Force Service Pilots (WASPs). Originally named Great Falls Army Air Base, later Great Falls Air Force Base, the facility was renamed Malmstrom Air Force Base on 1 October, 1955 in honour of Colonel Einar Axel Malmstrom. Malmstrom AFB is one of three U.S. Air Force Bases that maintains and operates the Minuteman III intercontinental ballistic missile. The museum portrays the base history from 1942 to 2012, covering the flying training mission and then the Minuteman Missile history from 1961 to present. The

The B-25 is now displayed at Hurlburt Field Memorial Air Park. She is seen here in 2016 and 2017. She has the five diagonal stripes of the 1st Air Commando Group in the CBI Theatre but has never flown there.
(Collection Wim Nijenhuis, Earl Leatherberry)

Malmstrom Museum includes both indoor displays and an outdoor park. Inside the museum is the largest collection of model military aircraft displays in the Northwest and a reconstruction of World War II barracks. Other parts of the museum include uniforms, a section of a module from an early Minuteman launch control centre, a cutaway of a Minuteman silo, and several displays that preserve the history and heritage of the base. The outdoor air park displays airplanes and ground transportation including airplanes from the World War II and the Vietnam era. One of these airplanes is the former movie star B-25. This is N9451Z. After storage at Davis Monthan, she was sold to National Metals Inc. at Phoenix, Arizona and registered as N9451Z. In August 1960, she was sold to Sprung Aviation and planned for conversion into a tanker. But the airplane was stored until July 1968, when she was purchased by Tallmantz Aviation at Santa Ana, California. She had a leading role in the film "Catch-22" and was later modified with a camera nose and flew with Tallmantz.Aviation. In 1986, the bomber was traded into the USAF Museum Program and delivered to the museum at Malmstrom. In July of that year, she was displayed on pylons near the main gate. The airplane flew in to Malmstrom's airfield earlier from Sherman Aircraft Sales, West Palm Beach, Florida before preliminary storage at Hangar one for the celebration in July. An upper turret was mounted on the fuselage and a solid gun nose was mounted on the airplane to again reflect a J model. Later the airplane was painted in an AAF Olive Drab and Neutral Grey scheme and with the squadron insignia of the famous Bridge Busters, the 490th Bomb Squadron, 341st Bomb Group.

In 1986, N9451Z was delivered to the USAF at Malmstrom AFB. A solid gun nose replaced the camera nose and the airplane was painted in an Olive Drab and Neutral Grey scheme. The airplane was put on display at the base air park. She has the markings of the 490th Bomb Squadron, 341st Bomb Group. (Jayson F. Snow, Montana Office of Tourism)

MAXWELL AFB, MONTGOMERY, ALABAMA

Maxwell Air Force Base (also known as Maxwell-Gunter Air Force Base) is located in Montgomery, Alabama. The Maxwell-Gunter base is home to the 42nd Air Base Wing (ABW) and headquarters of the Air University of the U.S. Air Force. It is named after the Second Lieutenant William C. Maxwell.

The B-25 at Maxwell AFB is N9452Z with the serial 44-30649. In 1968, she was sold to Tallmantz Aviation, Orange County for use in the film "Catch-22". In 1987, she was transferred to Maxwell and is displayed outdoors as "Poopsie".

"Poopsie" at Maxwell AFB in March 1992. She is displayed as #42–53373, faking a B–25C–5 model, but in fact she is #44–30649. (Simon Brooke)

"Poopsie" in June 2016. Unfortunately, the transparent glass surfaces are now sealed.

(Weslatham)

MUSEUM OF AVIATION HISTORY, WARNER-ROBINS, GEORGIA

The Museum of Aviation History is located at Robins Air Force Base in Warner-Robins, Georgia. It is situated on 51 acres adjacent to the base and is the second largest in the U.S. Air Force museum system. It has a total of five different buildings containing nearly 100 different airplanes. A vast array of his-

in the Department of Defence. Robins AFB served as a depot repair and supply support facility for B-25s in the southeast during World War II. It is, therefore, no wonder that a B-25 is included in the museum's collection. The B-25J on display is a composite of several airplanes to which the serial

acquired the airplane through an exchange in 1987. After storage at Davis Monthan, the B-25 was sold to Blue Mountain Air Service in October 1958 and registered as N2888G. In 1961 and 1962, the airplane was sold and from 1968 until 1983, reported derelict and stripped at Boise, Idaho. In 1983, she was acquired by the Pacific Museum of Flight. Aero Nostalgia Inc., Stockton, California bought the airplane in 1986. The hulk was trucked to Stockton and she was rebuilt for static display. In 1987, she was displayed at the museum at Robins AFB, and displayed as "The Little King".

torical airplanes, vehicles, engines, missiles, and other artefacts are displayed. The Museum of Aviation opened to the public on 9 November, 1984 with 20 airplanes on display in an open field and another 20 were in various stages of restoration. The Museum of Aviation has grown to become the second largest museum in the United States Air Force and the fourth most visited museum

number 44-86872 was assigned. The real "872" was delivered to the Army Air Corps in August 1945 and served in various units and throughout the United States until 1958, when she was retired. The airplane is currently marked as combat veteran #43-27676 "Little King" assigned to the 310th Bomb Group, and 380th Bomb Squadron serving in Europe. The Museum of Aviation

The B-25 with the tail colours of the 310th BG, 380th BS in April 2013 at the Museum of Aviation History. On the left fuselage nose the nose art and name "The Little King".

(Alan Wilson, Collection Wim Nijenhuis)

N61821 as the PBJ-1D of the National Naval Aviation Museum at Pensacola. The airplane had been stored outside for years, and continued exposure to the elements has resulted in significant corrosion. She is seen here at Pensacola storage in October 2014.

(John Tomlinson, igor113)

National Naval Aviation Museum, Pensacola, Florida

The National Naval Aviation Museum is the largest Naval Aviation museum in the world. It shows the Naval Aviation's rich history and has many restored airplanes representing Navy, Marine Corps, and Coast Guard Aviation. These historic and one-of-a-kind airplanes are displayed both inside the museum and outside. The museum is located at Naval Air Station Pensacola, Florida and it opened its doors on 8 June, 1963. It was housed in a renovated wood-frame building constructed during World War II. The growing collection at Pensacola quickly overwhelmed the capacity of the museum to display the airplanes. The museum needed to expand to meet the growing demands placed upon it. So, it expanded in 1975. New expansions were made in 1980 and 1990. The museum has evolved over the years into a steadily growing and expanding institution of national significance and is now considered the leading tourist attraction between Orlando and New Orleans and one of the top ten attractions in Florida. The National Naval Aviation Museum features over 150 beautifully restored airplanes, including rare and one-of-a-kind flying machines. One of those machines is a B-25J-15 with s/n 44-29035. She is displayed as a PBJ-1D. In 1959, the airplane was registered as N3516G. In the 1960s, she operated as a fire tanker by Paul Mantz Air Services. She even was flown to Caracas, Venezuela for a fire tanker contract and was later sold to the Venezuelan Air Force. In 1992, she returned to the U.S. and was registered as N61821. She was sold to Tony Mazzolini/U.S. Aviation Museum at Cleveland, Ohio in July 1993. She was trucked to Cleveland for static restoration for the National Naval Aviation Museum and was completed at Beaver Falls, Pennsylvania in March 1998. The B-25 went to the museum at Pensacola in March 1998. The PBJ on display was painted in the markings of airplane MB-22 assigned to VMB–423 that crashed on the island of

Vanuatu on 22 April, 1944. The museum's airplane had been stored outside for years, and continued exposure to the elements has resulted in significant corrosion. The machine was restored to commemorate the 75th Anniversary of the Doolittle Raid and is painted in the colours of the B-25B flown by its leader, Lt. Col. James "Jimmy" Doolittle. Over 4,000 hours have been invested in the restoration, much of which consisted of removing and repairing corroded areas and treating sections of the machine with a corrosion preventative. A great deal of time has also been spent ensuring that the markings perfectly match those seen on Doolittle's original machine. In April 2017, the National Naval Aviation Museum placed their newly restored B-25 on public display.

The machine was restored to commemorate the 75th Anniversary of the Doolittle Raid in April 2017 and is painted in the colours of the B-25B flown by its leader, Lt. Col. James "Jimmy" Doolittle.
(DougK)

Pacific Aviation Museum Pearl Harbor, Honolulu, Hawaii

One of today's B-25s in the colours of a Doolittle Raider is exhibited in the Pacific Aviation Museum Pearl Harbor, a very appropriate place for this airplane. The Pacific Aviation Museum Pearl Harbor is located on Ford Island (a secured military base) in the middle of Pearl Harbor. The Pacific Aviation Museum Pearl Harbor is a non-profit foundation founded in 1999 to develop an aviation museum in Hawaii.

The museum hosts a variety of aviation exhibits with a majority relating directly to the attack on Pearl Harbor and World War II. The idea for the Pacific Aviation Museum began on the anniversary of the Victory over Japan in 1995. Since first opening its doors to the public in December 2006, the Pacific Aviation Museum Pearl Harbor has grown steadily, adding aircraft and exhibits, now occupying two WWII hangars. Visitors to the Museum arrive via shuttle at Ford Island, a former Naval Air Station and National Historic Landmark located in the middle of Pearl Harbor. From the island's iconic, 158-foot control tower to the two World War II-era hangars that house the museum's air-

craft collection, signs of the surprise attack are still visible to this day. One hangar is Hangar 37, a 42,000 square foot former seaplane hangar that survived the December 1941 attack. The Hangar 37 opened with the museum on 7 December, 2006, and features much of the museum's static exhibits. The other is Hangar 79, a 80,000 square foot seaplane hangar. At each end, the hangar doors' blue glass windows are still riddled with bullet holes left by the Japanese attack. During the war, it was a maintenance and engine repair facility, filled with fighters, bombers and patrol aircraft that were based in Pearl Harbor or en route to the front lines. Today, it holds many of our modern jets and historic helicopters.

Originally, the B-25 of the Pacific Aviation Museum was a B-25J-30 with s/n 44-31504. In November 1951, she was taken on strength with the RCAF. By 1961 after her RCAF service, she was sold to Canspec Air Transport of Calgary, Alberta, Canada. In November 1961, she was sold to A.J. Warlick and J.E. Kowing at Seattle, Washington and registered as N9753Z. In May 1963, she was sold to Flair Inc. of Lihue, Kauai, Hawaii. In 1964, she was flown to Hawaii for the filming of "In Harm's Way". After filming, she was stored at Honolulu, Hawaii. By 1971, she was reportedly used by a technical school at the Honolulu Airport. She was obtained by a local Air National Guard unit and restored for display at Hickam Air Force Base. In 1976, she was painted in a red, white, and blue bicentennial colour scheme that would later be removed. In 2001, she was moved to Ford Island with the intention of being restored for the newly announced Pearl Harbor Museum. It was discovered that she was too heavily corroded that res-

toration would be impractical. At that time, she was traded to Aero Trader as partial payment for restoration of another B-25 that is currently on display at the Pearl Harbor Museum. This static B-25 is composed of several B-25 airplanes. The cockpit and tail are used from s/n 44-30077, the wings are from s/n 44-30627 and the centre section is from another airframe.

Today, the Pacific Aviation Museum at Ford Island has a B-25 in the colours of a Doolittle raider. After bouncing around for four decades, major parts of the B-25J, s/n 44-30077, were incorporated into the "mock" B-25B put together for the museum. In April 2007, the airplane was equipped with the nose art of the "Ruptured Duck". Hal Olsen, a retired naval aviation mechanic and professional nose artist, painted the "Ruptured Duck" on the nose of the bomber in honour of the 65th Anniversary of the Doolittle raid. (Collection Wim Nijenhuis)

THE FLYING LEATHERNECK AVIATION MUSEUM, SAN DIEGO, CALIFORNIA

One of the very few B-25 Mitchells in the colours of the U.S. Marines is displayed in The Flying Leatherneck Aviation Museum at San Diego, California. This Mitchell is a B-25J-30 with s/n 44-86727. Serving after the Second World War in Canada as Mitchel Mk.III

#5230 of the RCAF, this B-25 did odd jobs through the 1960s. In 1961, she was sold to Woods Body Shop of Lewistown, Montana and registered as N92875. She was sold a few times and eventually was acquired by the U.S. Marine Corps in 1976 and displayed

N92875 in the late 1960s. The airplane has been converted with a belly tank and spray bars below the wings.
(Collection Wim Nijenhuis)

Here she is seen at MCAS El Toro in 1996 and painted in three-tone USMC colours.
(Tom Tessier)

The airplane at her new home The Flying Leatherneck Aviation Museum at MCAS Miramar. The dark blue fuselage top has now weathered to a green colour.

Left: Behind the scenes, the B-25 during restoration at the museum's workshop in 2013.
(Skytamer, MCAS Miramar)

as USAF airplane at the USMC Museum at MCAS Quantico, Virginia. In 1987, she was displayed at MCAS El Toro, California and painted as a PBJ-1J. With the closure of that base and museum, the B-25 was moved to the Flying Leatherneck Museum at Miramar, near San Diego. She is displayed outside and marked as a PBJ-1J. The Flying Leatherneck Aviation Museum was established in 1989 on MCAS El Toro in Orange County, California. When Base Realignment closed El Toro in 1999, a group of retired Marines from San Diego, led by Major Generals Bob Butcher and Frank Lang, spearheaded the move of the airplanes and artefacts to MCAS Miramar. The Flying Leatherneck Aviation Museum at MCAS Miramar is the only museum in the world dedicated to United States Marine Corps aviation, with the largest and most complete collection of vintage aircraft flown by Marine pilots in the world. The museum has currently more than 30 airplanes on display and eight galleries of artefacts from World War I to present day. The Museum maintains and restores vintage aircraft used by the United States Marine Corps through work performed by employees and volunteers.

USS ALABAMA BATTLESHIP MEMORIAL PARK, MOBILE, ALABAMA

USS *Alabama* Battleship Memorial Park is a military history park and museum located on the western shore of Mobile Bay in Mobile, Alabama. It has a collection of notable airplanes and museum ships. One of the airplanes is a B-25 with the registration N9463Z. This is a B-25J-30 with s/n 44-31004. She was converted and used as a sprayer by Dothan Aviation Corp., Dothan, Alabama. In February 1974, she was donated to the Battleship Memorial Park at Mobile. She is displayed outdoors as an airplane of the 345th Bomb Group and named "Mary Alice". In 2022, she was moved inside the Aircraft Pavilion.

N9463Z at the USS Alabama Battleship Memorial Park in 2002 and 2017. The nose art was changed, and the name "Mary Alice II" became "Mary Alice III". (Sandra Scott, Robert/Jan)

In May 2019, the museum started restoration of the B-25. The solid gun nose was converted to a glass nose and the other glass openings were changed. All this with the aim of presenting the aircraft as a Doolittle Raider. In 2022, after the restoration was completed, the plane was moved inside the Aircraft Pavilion and join the museum's WWII exhibit. (Cap Pattison)

A schematic representation of Circle-Vision 360°.

(imagineeringdisney.com)

Circle-Vision 360°

Theatres that completely surrounded the audience with movie screens were extremely popular attractions at theme parks, world fairs and expositions after the war. When Disneyland, located in Anaheim, California, opened in July 1955 the Circle-Vision Theatre, it accommodated Circarama, featuring the film "Tour of the West", sponsored by American Motors. The name of this cinematic presentation was apparently inspired by a cinema format called Cinerama. For the Circarama format, eleven film screens formed a full circle around the audience, displaying scenes shot by eleven 16-millimetre film cameras that had been mounted in a circle on top of an American Motors automobile. In 1960, Bell Telephone took over the sponsorship and operation of the film attraction and the method of presentation no longer needed to emphasize the "car". After "Tour of the West", the next film using the original eleven-camera technique, "America the Beautiful", was first shown at the 1958 World's Fair in Brussels, and then came to Disneyland in June 1960, running through September 1966. The film was sponsored by The Bell System and it featured a tour across the United States visiting some historical sites and important places. Without the need to emphasize an automobile, the circular contraption of multiple cameras was freed to be attached to other vehicles, such as helicopters, to move about. About this time, the eleven-camera 16-millimetre method's days were numbered. The switch was made to use nine 35-millimetre cameras to capture the complete circular view, and it was presented in other places before Disneyland. This version of the surrounding experience was called Circle-Vision 360° for these presentations elsewhere.

AMERICAN FILM INDUSTRY

The film camera and the B-25 often formed a fascinating and successful duo and the warbird was an ideal aerial camera platform. (Aviation Filming)

After the war, film makers needed real airplanes as an aerial camera platform for aerial shootings. Many films which had required an aerial camera platform, numerous television productions and various commercials have been filmed and photographed from B-25 Mitchell bombers. In the U.S.A., the B-25 was used in countless Hollywood productions and commercial advertising for airlines and other aviation companies. The J-model of the B-25 Mitchell series enjoys limited use as a camera ship. For motion picture camera ship missions, the nose, tail, and the waist gunner positions serve as mounting positions for film cameras. The B-25 was an ideal platform from which to undertake aerial photography. Not only was it a good steady flight platform, but with the ending of the war and the large numbers of surplus B-25s now entering the civilian market, they were cheap to purchase. Paul Mantz later merged into Tallmantz Aviation, at Orange County in the U.S.A. was the leading company in aerial photography at the time. They modified a first B-25 for this purpose. Distinctive in the B-25's modification was the construction of a new camera nose. The camera noses had evolved over the years, beginning with a standard galvanized iron nose. With the coming of Cinerama, which required a wide-screen view for the bulky three-camera assembly, it became a distinctive wraparound nose glass. The wraparound nose glass was specially formed as a cylindrical section with minimal optical distortion for filming. It was flatter at the base and wider overall than the standard greenhouse nose and allowed the installation of a section of curved, optically clear, cylindrical wraparound nose. The waist and tail gun positions of the B-25 also were modified to accept camera mounts. The airplane was probably also the only plane suitable for the unique Disney Circle-Vision technique. This was a nine-camera assembly shooting 360 degrees of film. The assembly was tucked into the bomb bay and was lowered for filming. In this configuration the camera captured a full 360-degree field of view. When projected in the theatre, the visual field is recreated in all directions with just a one-inch gap between the nine screens. Disney's Circle-Vision 360° format set the standard for multi-screen, surround cinema back in the 1960's with the film "America the Beautiful". More recently, a B-25 of Chino's Plane of Fame Museum was fitted with a Spacecam gyro-stabilized ball mount in the tail gunner's position for "Con Air" to do air-to-air shooting of that film's C-123 Provider story ship. It was also used for the filming of "Pearl Harbor". For some time, the B-25 was the fixed-wing camera platform of choice, but that role has been largely taken over by the modern airplanes, helicopters, and drones.

FILMWAYS, INC., HOLLYWOOD, CALIFORNIA

Filmways, Inc. (also known as Filmways Pictures and Filmways Television) was a film and television production company founded by American film executive Martin Ransohoff and Edwin Kasper in 1952. Edwin Kasper left the firm in 1957. It is probably best remembered as the production company of CBS' "rural comedies" of the 1960s, including "Mister Ed", "The Beverly Hillbillies" and "Petticoat Junction", as well as the comedy-drama "The Trials of O'Brien", the western "Dundee and the Culhane", the adventure show "Bearcats!", the police drama "Cagney & Lacey", and "The Addams Family". Notable films the company produced include "The Sandpiper", "The Cincinnati Kid", "The Fearless Vampire Killers", "Ice Station Zebra", and Brian De Palma's "Dressed to Kill" and "Blow Out".

movies for them. Filmways housed studios in Manhattan, which were built for Metro-Goldwyn-Mayer in the 1920s. In 1974, Filmways acquired book publisher Grosset & Dunlap from American Financial Group. In May 1975, it bought the television syndication firm Rhodes Productions from Taft Broadcasting. In 1978, it acquired Ruby-Spears Productions, which had launched a year earlier. In 1979, Filmways purchased American International Pictures. Their TV subsidiary, AITV, became a Filmways' new syndication division in 1980, spinning off Rhodes into an independent corporation. Filmways had lost nearly $20 million during the nine months ending in November 1981. However, it partially exited bankruptcy by selling a few of its previously acquired assets. In 1982, Filmways was acquired by

camera ships. Later, an additional ship was bought by Tallmantz in Mexico. After filming in early 1971, the planes were parked on the Tallmantz ramp and had to be sold. Tallmantz acted as sales agent, and ads were posted offering the airplanes. The B-25s were gradually sold and by 1976 they were all sold. The B-25s of Filmways were all purchased by Tallmantz in 1968 and had the following registrations:
N10V, N9856C, N7687C, N9115Z, N10564, N2849G, N9451Z, N9452Z, N8195H, N3699G, N9494Z, N3174G, N7681C and N3507G. In 1978, three of these B-25s were used in the United Kingdom for the film "Hanover Street". These were N7681C, N9115Z and N9494Z.

More information and pictures of the B-25s of Filmways are shown in the parts about the relevant films and organisations elsewhere in this book.

The wraparound nose of the first B-25 camera ship of Tallmantz Aviation N1203.
(Collection Wim Nijenhuis)

The firm listed on the New York Stock Exchange in 1959. Filmways started making TV commercials, moved into documentaries then sitcoms. In 1962, Filmways produced its biggest hit, "The Beverly Hillbillies" for CBS. In 1966, the company acquired Heatter-Quigley Productions, the game show producer known for their biggest hit "Hollywood Squares". In 1969, it bought Sears Point Raceway in Sonoma County, California and Wally Heider Recording with studios in Hollywood and San Francisco, along with Studio 3 Inc. in Hollywood. In 1972, Ransohoff left Filmways as president and became an independent producer. He signed a contract with Paramount to make

Orion Pictures. Filmways was then reincorporated as Orion Pictures Corporation on 31 August, 1982. Today, most of the Filmways library is owned by MGM, successor-in-interest to Orion which it purchased in 1998, until Orion Pictures was quietly relaunched by MGM on 11 September, 2014.

In relation to the B-25, the best-known film of Filmways was "Catch-22". This film was directed by Mike Nichols, released by Paramount Pictures in 1970 and produced by Filmways. On behalf of Paramount Studios and subsidiary Filmways, Tallmantz Aviation located and purchased fourteen B-25s. Tallmantz owned three B-25s, including two

TALLMANTZ AVIATION INC., SANTA ANA, CALIFORNIA

Tallmantz Aviation had been supplying airplanes and aircrews for motion pictures for a long time and displayed their vintage airplane collection at the Movieland of the Air Museum. Tallmantz Aviation owned a trio of B-25 film camera platforms. Tallmantz Aviation was based at Orange County Airport at Santa Ana, California, now John Wayne Airport. Tallmantz Aviation supplied pilots, camera planes and a small fleet of antique and historic airplanes for film and television productions. Tallmantz Aviation was incorporated in November 1961. The Board of Directors was made up of Franklin K. Kline III, Frank G. Tallman, and A. Paul Mantz. Records mention Frank Tallman as Corporation President and Paul Mantz as Vice-President. Mantz brought with him his specially modified B-25 (N1203). The firm soon purchased a second B-25 to handle the increased

The driving forces behind Tallmantz Aviation Inc.

Left: *Paul Mantz in the mid-1950s with his famous camera ship N1203. Mantz was Vice-President of Tallmantz Aviation. He spent a good deal of time flying across the world in this converted bomber. The map on the side of the fuselage documented the world-ranging trips flown by Mantz. (SDASM)*

Frank Tallman was Corporation President of Tallmantz Aviation. After the death of Mantz in 1965, the company carried on with Tallman at the helm. In the picture, Tallman in front of Tallmantz' second camera ship N1042B. (Collection Wim Nijenhuis)

Frank Pine, in front of N1042B, was long time chief pilot/general manager for Tallmantz. He became Corporation President after Tallman's death in 1978. (Aerovintage.com via James Rogers)

workload. Since 1961, Tallmantz had modified a third B-25 and a Douglas A-26 Invader for filming. For film work demanding a slower camera platform, they had a Curtiss Junior from 1930 and a Stinson L-1 from 1939. Unfortunately, Mantz's and Tallman's collaboration did not last long. In 1965, the two men were working on the film "The Flight of the Phoenix" when Tallman, who was supposed to perform a landing sequence in the Arizona desert, shattered his kneecap during a fall at home. Mantz took his place and on 8 July, Mantz was performing the landing when one of his airplane's wheels hit a small mound of sand and caused him to lose control. The airplane crashed and

Mantz was killed instantly. A few days later Tallman faced his own personal tragedy when doctors amputated his leg because of a massive infection that had resulted from his broken kneecap. Despite the loss of his leg and of his close friend, Tallman taught himself to fly using only one leg and returned to stunting.

After the loss of Mantz, the company carried on with Tallman at the helm. Frank Pine became Vice President and also chief pilot. Frank Pine had been a long-time chief pilot/general manager for Tallmantz. He flew almost all the B-25 camera ships for Tallmantz. Martha Marchak, Mantz's long-time secretary, was both corporate secretary and treasurer and became later Pine's wife. On 15 April, 1978, Tallman lost his life during a routine flight when he failed to clear a ridge

near Palm Springs, California, due to poor visibility. With the loss of Tallman, Pine became the Corporate President. But only six years later, in 1984, Pine died following a cardial attack. With the loss of Pine, the days of Tallmantz Aviation were numbered. The last filing for the company, dated October 1984, showed Walter W. Pine (Frank's brother) as Chief Executive Officer and Martha Marchak Pine as secretary and Chief Financial Officer. The Board of Directors consisted of Martha Marchak Pine and Ruth Marchak Tallman. The company was dissolved by both directors, a process that began in February 1985 and ended on 30 January, 1986. The company was sold to outside investors who reoriented the company completely from its cinematic past. The large airplane collection was sold and the two B-25s went to Sherman Airplane Sales, an airplane brokerage firm in Florida.

THE TALLMANTZ CAMERA SHIPS

Probably the most famous civilian Mitchells were those of Tallmantz. The company owned a trio of B-25 film camera platforms. Throughout the heydays of Tallmantz Aviation, from the early 1960s through the mid-1970s, these B-25s were a well-known sight all over the world. The three B-25s were the most successful airplanes modified for aerial filming and were utilised for film work in many motion pictures, television, and commercial productions. The three B-25s had the civil registrations N1203, N1042B and N9451Z.

N1203

This was a B-25H-5 with s/n 43-4643, Mantz selected as a first camera ship. She was delivered in March 1944 and assigned to domestic units during the war. She was declared surplus in October 1945 and sent off to Stillwater for disposal. Mantz registered the airplane on 28 May, 1946 as NX1203 and flew her to Grand Central Air Terminal at Glendale, California. She was put to work quickly by Mantz, reportedly being used in the summer of 1946 to film shots that appeared in the film "Best Years of Our Lives". She was modified to accept camera mounts in the tail, waist, and nose, though the early nose shots were evidently made through the removed emergency exit hatch located there. After this beginning, Mantz put the B-25 to good use in many post-war film productions in the first ten years after the war. In 1954, with the coming of Cinerama, which required a wide-screen view

for the bulky three-camera assembly, Mantz modified the B-25J nose he had mounted to accept a special nose glass piece that offered a panoramic distortion-free view. It was flatter at the base and wider overall than the standard greenhouse nose and allowed the installation of a section of curved, optically clear, cylindrical wraparound nose. A large rotating mounting plate, built up from spare parts of a B-17 Sperry ball turret, was fitted at the base of the nose section to

Left: The early Cinerama wraparound nose. The camera glass is protected by metal rolling doors, an early effort to protect the camera view from bug splatters prior to filming. The large maps adorning both sides of the nose document the airplane's world-spanning film missions. Mantz added the nose maps during the Cinerama missions, with national flags bordering the map. Route tracings were added to show where the airplane had flown. (SDASM)

This picture shows N1203 in 1954 in the early colours of Paul Mantz at Orange County Airport. The camera nose at this point is a modified standard B-25J nose. **Left:** *The same ship around 1957. (Dusty Carter, WIX)*

accept the Cinerama cameras. On the upper portion of the nose there were large hatches allowing the positioning and installation of the equipment. Smaller windows were added around the wraparound glass which increased visibility for the camera operator. A problem was that in practice the beautiful glass nose was quickly covered with splattered bugs during the low-level work. This problem caused the airplane to be dubbed "The Smasher". Other camera positions were mounted in the fuselage, tail and two waist blisters. The tail mount was constructed by removing the extreme aft portion of the tail and installing the camera, providing

an unlimited view to the rear. This position was open, so no glass was needed. When the camera was not mounted, the normal tail was installed. A Plexiglas door in the interior near the rear bulkhead and a gas-powered heater provided a warm environment for the crew when the rear position was manned. The two waist mounts provide limited, but sometimes useful camera angles. The left position was obtained by removing the Plexiglas window and shooting through the opening. The right position was obtained by removing the circular emergency hatch in the rear of the fuselage. The airplane was painted in an overall two-tone grey and white scheme with red engine cowlings and nacelles and the twin tails in red. The airplane registration number was painted vertically on the stabilisers along with the U.S. flag. Later a world map would grace each side of the fuselage below the cockpit, detailing the many trips she had made with the flags of the nations she had visited.

The first Cinerama film, "This is Cinerama", which was released in 1952, had Mantz flying his B-25 at low level on a U.S. tour. Over the next decade, Mantz became the Cinerama pilot and continued work for subsequent productions. By 1962, Mantz had long perfected his camera nose and it perfectly suited the camerawork required for the Cinerama filming. In August 1956, the airplane was transferred to Paul Mantz Air Services. When Walt Disney decided to film a feature using an eleven-camera 360 degree process, initially dubbed Circarama (later Circle-Vision) in 1957, Mantz and his B-25 were called upon to carry the heavy camera and, through a complex mount, suspend it from beneath the bomb bay of the bomber during aerial filming. She was equipped with the Circle-Vision camera assembly suspended from the bomb bay for aerial photography in 1958. The airplane was transferred to Mantz Air in November 1960 and transferred to Tallmantz Aviation in November 1961 and registered as N1203. As mentioned before, Mantz got killed in July 1965, but his prized camera ship continued with Tallmantz and, in fact, was used to carry Mantz's mortal remains back from Yuma after the accident that took his life. In 1969, after the film work for "Catch-22" had

The left side map was slightly changed in 1958. The information contained inside the border of the map area was deleted and moved aft of the flags surrounding the map. The markings on the nose that say "Darryl F. Zanuck Productions, Inc., 20th Century-Fox Special", was a 1956 Mantz project. (Collection Wim Nijenhuis)

been completed, N1203 returned to her Orange County base. In the following years she continued with Tallmantz but Frank Tallman made a decision that put operational considerations above sentimentality. The old Mantz B-25H had not had any significant system upgrades since the 1940s and Mantz, in an effort to make the interior of the B-25 quieter for his film-making companions, had covered the inside walls with thick sound insulation. All that made N1203 a mechanic's nightmare. Tallman decided to retire N1203 in favour of a B-25N, N9451Z, a "Catch-22" veteran. By 1971, the camera nose was removed from N1203 and, for the first time in a quarter century, a standard B-25J nose was mounted. In July 1975, N1203 was sold to Howard Stucky and Lawrence Leang of Moundridge, Kansas. In March 1976, the airplane was sold to Leroy Sansom of Burbank, California, but quickly sold the same month to Vicki Meller, also of Burbank. She was last reported seen at Van Nuys Airport, still in the Tallmantz paint scheme, but

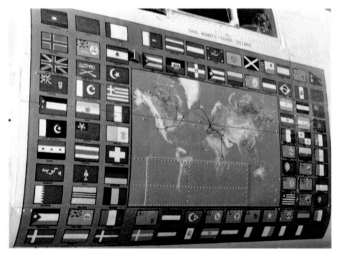

N1203 in the mid-1960s. At right, a detail of the world map on the left side of the nose. (Nick Williams, AAHS)

carrying the name Talisman Aviation on the nose. She was being watched for drug runs conducted south of the border, Tallmantz Aviation at one point being contacted by anti-drug agents about their old airplane. But the airplane ultimately disappeared after 1976. No further information has surfaced about her fate other than a rumour that she had crashed in Colombia sometime later. Her fate is unknown and remains a mystery.

N1042B

The second camera ship of Tallmantz was the B-25J-25 with serial number 44-30823. In 1958, this airplane was sold to the civil market and was purchased by Wenatchee Air Services, Yakima, Washington. She was registered as N1042B. In May 1962, she was sold to Tallmantz Aviation. When Tallmantz Aviation was incorporated in November 1961, one of the first tasks undertaken by the new company was to find a second B-25 for use as a camera ship, supplementing Mantz's original N1203. With N1042B, they found this airplane. The airplane was overhauled as needed and modified similarly to N1203 for use as a camera ship. The airplane was nicknamed "Lively Lady". She

lage. Extensive lettering was added to the nose, identifying the airplane's operator, as well as a world map (duplicating that on N1203) with a list of projects the airplane was involved in. The airplane was used from 1962 until 1985 as a Tallmantz camera ship and flew to many different places around the world.

This view shows N1203 with the new paint scheme at Orange County in the 1970s. She was sold in July 1975. The airplane went through the hands of several owners in the next few years and her most likely fate was a crash in Colombia during an abortive drug run.
(Collection Wim Nijenhuis)

Tallmantz did numerous Circle-Vision 360 projects for Walt Disney in the 1960s and early 1970s for which N1042B was used with Frank Pine as the pilot. The Circle-Vision camera assembly weighed several hundred pounds and involved nine cameras filming simultaneously on a mount that was lowered from the bomb bay of the bomber. In 1969, she flew in a desert cam-

was converted into a cinema camera ship, with a camera in the nose and tail, and a camera gantry in the bomb bay that could be extended and retracted in flight. Distinctive in its modification was the construction of a new camera nose. The nose of N1042B was constructed from scratch, was slightly longer than the nose of N1203, and incorporated a large hatch on the left side to assist in mounting cameras. The airplane was finished in an overall white scheme with red rudders and wingtips, and a black anti-glare panel in front of the cockpit. The engine nacelles were blue. A thin black cheat line extended from the cockpit to the aft fuse-

ouflage pattern in the film "Catch-22" and in 1971 after filming, she was repainted overall white with blue engine nacelles and received the distinctive Tallmantz emblem on the vertical stabilisers. Also, the nose markings were updated. A half-bubble dome for the aerial director's position was fitted in the former top turret skin patch, wired into the radio intercom system. A comfortable, elevated seat was provided for the aerial director, who could communicate instantly with any of the film crew trough an intercom system while observing the actions. The director was also able to communicate directly with other airplanes involved in the

Left: Early view of N1042B when owned by Tallmantz with the wraparound nose, but without the Tallmantz name on the nose. (Collection Wim Nijenhuis)

Right: An early paint scheme was used for expo67. The markings on the nose representing the Disney Circle-Vision 360 project for expo67 held at Montreal, Canada. The 1967 International and Universal Exposition is the proper name, but for short it is expo67. It was a World's Fair held in Montreal Canada in 1967.

(Collection Wim Nijenhuis)

filming by special radio installation. The airplane has participated in over 80 Hollywood films. When Tallmantz was sold to new owners in 1986, the airplane was sold to Universal Aviation, a corporation in Delaware. The airplane was operated by Aces High Ltd. in the United Kingdom. Aces High also used the B-25 as a camera ship and she was seen on several air shows in Europe. Further described in the chapter United Kingdom/Aces High. In 1996, the airplane returned to the U.S.A. after she was sold to World Jet Inc. at Fort Lauderdale, Florida. There she was overhauled by Tom Reilly and continued to fly until today. Nowadays she flies with the name "God and Country".

N9451Z

This airplane, a B-25J-25 with s/n 44-30493, was stored at Davis Monthan in 1958. In January 1960, she was sold to National Metals Inc. at Phoenix, Arizona and registered as N9451Z and in August the same year, sold to Sprung Aviation. She was slated for conversion to an air tanker until four B-25 tankers crashed in July 1960 while fighting fires. In the fallout, the B-25 design was undeservedly blamed for the crashes, the result of pilot error and mechanical failures caused by poor maintenance. As a result of one single event, dozens of B-25s purchased for air tanker duty were pretty much useless, and N9451Z languished at Tucson for the next eight years. In 1968, Tallmantz Aviation

N1042B was repainted in the early 1970s. The Tallmantz emblem was added to the vertical tail surfaces and the nose markings were altered. Painted on the nose are the symbol and information of the 1972 Disney Circle-Vison project. (Martin J. Simpson)

Nice close ups of the camera nose and tail, photographed at Oxnard Airport, California in the early 1970's. (Doug Duncan)

N1042B in action at Old Warden Aerodrome, Bedfordshire, England in 1972. The nine-camera assembly was mounted in the bomb bay with a hydraulically actuated trapeze assembly that was lowered for clear-view 360-degree filming. Note the attitude of the B-25 vs that of the camera. It is probable that the relatively low speeds used for the filming required the B-25 nose-high attitude and the camera mount was adjusted accordingly. (AJCDuxman)

began quietly looking for B-25s on behalf of Paramount Studios for use in the filming of "Catch-22". In July 1968, they purchased N9451Z and ferried the airplane back to Orange County. N9451Z was named "Dumbo" and was the airplane star of "Catch-22". This was the centrepiece airplane featured in many scenes with actor Alan Arkin as Capt. John Yossarian. After filming she was sold back to Tallmantz in August 1971. Tallmantz decided to replace N1203 with the updated N9451Z. The airplane was pulled into the Tallmantz maintenance hangar and stripped down, overhauled, and modified as a camera ship. The Cinerama nose came from N1203 and was mounted on N9451Z. The new camera ship was repainted in a glossy white scheme to match the new scheme used on N1042B, the other B-25 of Tallmantz. The paint scheme was complete with the new red, blue and black Tallmantz emblem on the stabilisers and she was

dubbed "Marty" in honour of Martha Pine. Shortly afterwards, N9451Z joined the flight line. Her first big project was the Seekval program for Boeing in 1974. Seekval was basically a government evaluation of an all-service simulator training technique then under development. The airplane would make low level runs across Army gunnery ranges in Arkansas and Washington. The cameras were positioned in the tail. It embarked on numerous film and military projects in the next decade, regaining her old standard B-25J nose at least once to allow higher airspeeds in test programs than allowed with the Cinerama nose. After the death of Frank Pine in 1984, Tallmantz Aviation was sold to new owners. N9451Z was placed up for sale, brokered by Sherman Aircraft Sales, but was ultimately traded into the USAF Museum Program in 1986. An upper turret was mounted on the fuselage and a solid gun nose was mounted on the airplane to again reflect a J model and the B-25 was painted in an USAAF colour scheme and displayed on pylons at Malmstrom AFB, Great Falls, Montana.

A picture of N9451Z at Santa Ana. She still has the colours used in the film "Catch-22" and the picture was probably taken shortly after the return from filming in Mexico. (Hans Melin)

In 1974, the ship was used for the Seekval program. Project IC1 Seekval (Imagery Collection Part I) was a limited test to define and attempt to resolve the technical, mechanical, logistical, and administrative problems involved in the collection of photographic and infrared (IR) imagery. The overall purpose of Project IC1 was to ensure the adequacy of the hardware, flight profiles, tactics and instrumentation to be used in the comprehensive collection effort, and if possible, to supply imagery for observer evaluation on the MMS (Multi-Mission Simulator). (Martin J Simpson)

Tallmantz modified the airplane as a camera ship to replace his old B-25H N1203. She was overhauled and the special camera nose from N1203 was mounted. The modified B-25 camera ship is seen here on the Tallmantz ramp in October 1979. (Steve Williams)

Picture of the camera nose of N9451Z at the Tallmantz ramp in February 1980. On the nose an emblem of the Seekval program. (Scott Thompson)

TALLMANTZ AND "CATCH-22"

The motion picture "Catch-22" has become renowned for its role in saving the B-25 Mitchell from a possible extinction. The film was made in 1969 and is a satirical war film adapted from the book of the same name by Joseph Heller. The film was directed by Mike Nichols and released by Paramount Pictures in 1970. The leading role in the film has Captain John Yossarian, a fictional character. A 28-year-old captain and bombardier in the 256th Bombardment Squadron of the Army Air Corps, stationed on the small island of Pianosa off the Italian mainland during World War II. Yossarian is played by the American actor Alan Arkin.

Joseph Heller enlisted in the Army Air Corps in 1942 and was sent to Corsica two years later. While in the Mediterranean Theatre, he flew 60 combat missions with the 488th Bomb Squadron, 340th Bomb Group of the 12th Air Force. Heller wrote a satirical anti-war novel about the "lunatic characters" of a bombardment group in the Mediterranean Theatre of Operations in the Second World War, which was published in 1961. Paramount Pictures assigned a $17 million budget to produce the film. In the opening scene sixteen B-25s taxi out to the runway and make a mass take-off. For the aviation enthusiast, this is one of the most impressive scenes of the film and it has never been repeated in any other film. Paramount planned to film the "Catch-22" aerial se-

From left to right, actors Paula Prentiss, Alan Arkin, and Collin Wilcox with director Mike Nichols in front of one of the bombers during the filming of "Catch-22". In the film, a total of 18 B-25s were used, 17 were flyable and an additional B-25 was acquired in Mexico, made barely ferriable and flown with the landing gear down to the filming location, only to be burned and destroyed in the landing crash scene. (Julian Wasser)

quences in six weeks, but the production required three months to shoot, and the bombers flew a total of about 1,500 hours. They would appear on screen just for about 12 minutes. Paramount constructed an airstrip location in San Carlos, Sonora, Mexico. They built a "base" on the site with a hospital, a large brick house, mess hall, control tower, bomb dump and enough pyramid tents to house the entire cast and crew. The location was detailed right down and looks very realistic as a Mediterranean air base during World War II. They constructed a 6,000' X 200' runway, with a perimeter taxiway and a hardstand for each airplane.

Frank Tallman and his crew of pilots at San Carlos. For the film, Tallman was primarily looking for military trained pilots who were familiar with formation flying. He completed each flight crew with a co-pilot and a flight engineer. These crews stayed together for the entire filming period. The flight crews received a week-long flight and ground school training. They were trained to fly in tight formations.
Right: *B-25s on the runway of the bomber base at San Carlos in Mexico. (Aerovintage.com, Chris Henry)*

The "Catch-22" budget could only afford seventeen flyable B-25 Mitchells. On behalf of Paramount Studios Tallmantz Aviation located and purchased fourteen B-25s. Tallmantz owned three B-25s, including two camera ships. One non-flyable ship was later acquired in Mexico. Of the eighteen B-25s in "Catch-22", two were B-25Hs and sixteen were B-25Js. When Tallmantz Aviation got the contract in 1967, they had to go out and find all these airplanes as well as a lot of spare parts to operate the fleet. In the middle of 1968, they started with the search for the airplanes. The airplanes came

The "Catch-22" airplanes

Emblem 488th BS, 340th BG

USAAF serial nr	Model	Civil registration	Catch-22 tail code	Film name
43-4432	B-25H-5	N10V	6N	Berlin Express
43-4643	B-25H-5	N1203	6A	
43-28204	B-25J-10	N9856C	6G	Booby Trap
44-28925	B-25J-15	N7687C	6F	Superman
44-29366	B-25J-20	N9115Z	6M	Hot Pants
44-29887	B-25J-20	N10564	6Y	Luscious Lulu
44-29939	B-25J-25	N9456Z	6C	
44-30077	B-25J-25	N2849G	6Q	The Denver Dumper
44-30493	B-25J-25	N9451Z	6V	Dumbo
44-30649	B-25J-25	N9452Z	6V / 6W	
44-30748	B-25J-25	N8195H	6H	
44-30801	B-25J-25	N3699G	6K	Vestal Virgin
44-30823	B-25J-25	N1042B	6E	
44-30925	B-25J-30	N9494Z	6P	Abominable Snow Man / Laden Maiden
44-31032	B-25J-30	N3174G	6D	Free, Fast and Ready
44-86701	B-25J-30	N7681C	6J	Annzas
44-86843	B-25J-30	N3507G	6B	Passionate Paulette
45-8843	B-25J-35	XB-HEY	6F / 6S	

S/n 43-4432, N10V. Appeared in the film as "Berlin Express" and had tail code 6N. She also featured in one sequence as a pristine VIP transport for Gen. Dreedle, complete with white sidewall tires, highly polished in an overall pinkish themed paint-scheme. In September 1968, the B-25 was bought by Tallmantz from Long Island Airways of Ronkonkoma, New York and sold to Dr. William S. Cooper of Merced, California in May 1971 who donated her to the EAA Air Museum Foundation.

(William T. Larkins, Collection Wim Nijenhuis, Stoney Stonich)

S/n 43-4643, N1203. She was used as a camera ship for the filming. She had no name but a squadron insignia on her nose. Her tail code was 6A. The airplane had the upper turret installed after the wings to represent an early B-25C or D type model. She was bought by Paul Mantz in 1946. In November 1961, she was sold to Tallmantz Aviation and in July 1975, sold to Howard Stucky and Lawrence Leang of Moundridge, Kansas. (Aerovintage.com)

S/n 44-29366, N9115Z. This ship flew in the film as "Hot Pants" with the tail code 6M. She was a fire tanker of I.N. Burchinal, Paris, Texas and was sold to Filmways in 1968. In 1972, after filming, she went to David Tallichet at Long Beach, California. (Stoney Stonich)

S/n 43-28204, N9856C. She flew in the film as "Booby Trap" and had tail code 6G. In September 1968, the bomber was purchased from Dennis G. Smilanch, Boise, Idaho for use in "Catch-22". In April 1971, she was sold to Tallmantz Aviation and in May of 1973, she was sold to Ted Itano, Borrego Springs, California.

(Stoney Stonich, Paramount)

from as far away as Long Island, New York. Others came from Grey Bull Wyoming, Buckeye Arizona, Houston Texas, Champaign Illinois, and forest fire tanker bases all over the western U.S. One of the things that was important for Tallmantz was that all the planes had the upgrade from Hayes in the 1950s. This upgrade concerned wiring, hydraulics and other systems. If this upgrade had not been done, then the airplane was not suit-

able because there was too much money needed for this upgrading of a B-25. They also had to be made ferriable to get them to southern California. Most of the B-25s were purchased in the second half of 1968 for under $6,000 each. In the fall of 1968, the airplanes arrived at Orange County. In the

hangar of Tallmantz, the systems of each airplane were checked and overhauled as needed. All the external civil modifications were removed and for the film mock upper turrets were installed to represent different models. Three airplanes had the turrets installed after the wings to represent early B-25C and B-25D models. The other planes had them before the wings, as was normal with the B-25J-models. Initially, the camera ships also had the mock turrets installed, but problems with buffeting necessitated their removal. The bomb bay doors were made operative. After the mechanical work

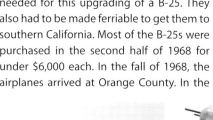

S/n 44-28925, N7687C. In August 1964, she was sold to Aerial Applicators Inc. and operated by Trans-West Air Service. In 1968, she was purchased for use in "Catch-22". During filming, she was painted in two colour schemes. The first was the basic olive drab and named "Superman" with tail code 6F. She also appeared in the film in the white colours of M & M Air Force. The picture at left was taken in 1969, after the makeover for the filming.
(Harry Gann via Norm Taylor via Milo Peltzer Collecton, Stoney Stonich)

S/n 44-29887, N10564. In October 1968, this B-25 was sold by Hemet Valley Flying Service in Hemet Valley, California and flew in the film as "Luscious Lulu" with the tail code 6Y. In January 1971, she was sold to David Allen of Davu Aviation Inc., Santa Fe, New Mexico. (Stoney Stonich, Charles Rector)

S/n 44-29939, N9456Z. In the 1960s, Paul Mantz owned this airplane and in November 1960, she was sold to Paul Mantz Air Services where she would be used as freight support for many movies. In November 1961, she was transferred to Tallmantz Aviation and was used in the film. She had no name but only a squadron insignia on her nose and the tail code 6C. In January 1971, she was sold to Donald Buchele of Columbus Station, Ohio. In the picture above, she is parked on the Tall- mantz ramp in September 1969.

(Stoney Stonich, Kenneth Johnson)

S/n 44-30077, N2849G. This B-25 was sold by Christler and Avery Aviation at Greybull in September 1960 and trans- ferred to Avery Aviation Inc., Greybull, Wyoming in August 1961. In 1968, the airplane was sold to Filmways and flew in the film as "The Denver Dumper" with the tail code 6Q. In August 1971, she was sold to Tallmantz Aviation and in December 1971, the airplane was sold to Keith Lar- kin & Peter Bell at Freedom, California.

(Collection Wim Nijenhuis, Chris Brame)

the airplanes received the authentic AAF camouflage schemes and markings. By the end of December 1968, the seventeen B-25s were ready for the departure to Mexico.

While in Mexico, each airplane received a tail code like in the real-World War II squad-

ron. The codes were painted in white figures on the upper part of the vertical stabilisers. The code was a number 6 and a letter suffix for each individual airplane. In contrast to the 488th squadron patch used on the fly- ing jackets; the tail code 6 was used by the 486th Bomb Squadron during World War II. The 488th Bomb Squadron used the num- ber 8. The top of the rudders was painted blue. The special effects department made the airplanes look war weary. Oil and paint were splashed over the airplanes and nose art was sanded to make it look like they had flown many missions. Studio artists applied nose art with names and other markings to make them look like real warplanes. In Mexico also the N of the registration was replaced with a 0, 1, 3 or 4. In a later period

*S/n 44-30493, N9451Z. In July 1968, this B-25 was purchased from Sprung Aviation, Tucson, Arizona. She was the centrepiece airplane featured in many scenes. She was named "Dumbo" and had the tail code 6V. The airplane had the upper turret installed after the wings to represent an early B-25C or D model type. After filming, she was modified with a camera nose and flew with Tallmantz. Aviation. In 1986, the bomber was traded into the USAF Museum Program and delivered to the museum at Malmstrom. **Below:** A picture with actors Alan Arkin in the nose and Art Garfunkel as co-pilot in the cockpit. (Chris Henry, Tampa Bay Times)*

eight or nine of the B-25s were repainted in a mottled light grey paint scheme with M&M markings while in Mexico. This was needed for filming a night attack on their own bomber base by order of M&M Enterprises from Milo Minderbinder. These airplanes carried no serial numbers or civil registration and cannot easily be identified.

After the completion of filming in May 1969, seventeen B-25s returned to Orange County in California. One airplane was destroyed in the film and was buried in San Carlos. The planes were parked on the Tallmantz ramp. Early in 1971, after having completed the film, Paramount asked Tallmantz Aviation to sell the airplanes for them. The prices ranged from $5,000 to $10,000 each, at the time this price seemed relatively high. In August 1971, Paramount asked Tallmantz to purchase the remaining airplanes. Tallmantz agreed reluctantly and finally the B-25s were sold by 1976. Tallmantz retained airplane N9451Z to be modified as a camera ship to replace the N1203.

S/n 44–30649, N9452Z. This B-25 was bought by Tallmantz from Don Underwood of Donaire Inc. at Phoenix, Arizona in June 1968. She flew in the film without a name and two tail codes, 6V and 6W. She had the upper turret installed after the wings to represent an early B-25C or D model type. In 1972, after filming, she went to the USAF at Wright–Patterson AFB, Dayton, Ohio. (Stoney Stonich, Todd Hackbarth)

S/n 44-30748, N8195H. In late 1969, her owner Christler and Avery Aviation of Greybull Wyoming sold her to Filmways for use in the film. During the film, she was assigned the tail code 6H. She carried no name or nose art, just the squadron emblem. Later in the film, she also appeared as one of the white "M&M Air Force" B-25s. In March 1972, she flew from Santa Ana to Van Nuys, California, still painted in the white "M&M" film scheme. "Miss Renee" was painted on for the ferry flight only. In that month, she was sold to Milan S. Pupich, George S. Pupich and Stephen H Crowe, Jr. and restored to airworthy condition.

(Stoney Stonich, Collection Wim Nijenhuis)

S/n 44-30823, N1042B. *This bomber flew in the film with the tail code 6E with squadron emblem and had no name or nose art. This was the second camera ship of Tallmantz, purchased in May 1962. After filming, she stayed with Tallmantz until 1986, when she was sold to Universal Aviation, Delaware. Later, she was operated by Aces High Ltd. in the United Kingdom.* (Aerovintage.com)

S/n 44-30801, N3699G. In February 1969, Christler and Avery Aviation of Greybull, Wyoming sold this B-25 to Filmways for use in the film. There she flew as "Vestal Virgin" with the tail code 6K. In August 1971, she was sold to Tallmantz Aviation. Challenge Publications of Canoga Park, California purchased her in February 1972 and registered her as N30801. (Stoney Stonich, Chris Henry)

Left & middle: *S/n 44-30925, N9494Z. In December 1968, owner John C. Estes of Beaumont, Texas sold this airplane to Filmways. She flew in "Catch-22" as "The Abombinable Snow Man" and later as "Laden Maiden". She had the tail code 6P. After filming, she was sold to the Confederate Air Force at Harlingen, Texas in February 1970. In the late 1970s, she was used as leading film star in the film "Hanover Street".* (Stoney Stonich)

S/n 44-31032, N3174G. This B-25 was sold by Walston Aviation Inc., East Alton, Illinois in September 1968 and flew in the film with the tail code 6D and was named "Free, Fast, Ready".

(Stoney Stonich, Collection Wim Nijenhuis)

S/n 44-86701, N7681C. In 1968, this B-25 was purchased by Tallmantz from Bud Marquis of Airplane Crop Dusting, Marysville, California. She was used in the film and flew as "Annzas" with the tail code 6J. In April 1978, she was sold to David C. Tallichet Military Aircraft Restoration Group at Chino and ferried to Luton for use in film "Hanover Street". At right, this airplane is unmarked, as were all the M&M B-25s, but it is believed that this is N7681C. Clearly visible is the mottled light grey white paint scheme. *(Jack Cook, J.D. Davis)*

S/n 44-86843, N3507G. This B-25 flew as a tanker and was sold by Donald Cochran, Montclair, California in May 1968 to Tallmantz. She was used in the film and named "Passionate Paulette" with the tail code 6B. After filming, the airplane was donated to Grissom AFB in 1972. There she was also displayed as "Passionate Paulette". *(Stoney Stonich, Dave Welch)*

S/n 45-8843, XB-HEY. This B-25, registered as N8091H, was exported to Mexico, and became XB-HEY. She was located by Tallmantz at an airfield near Guyamas, Mexico and owned by Charles Rector. In January 1969, she was purchased by Tallmantz for use in the film. She had no name and the tail codes 6F and 6S. This B-25 did not fly for the film but was used for a sequence depicting a B-25 crash. The burnt remains of the airplane were buried on the airstrip site at San Carlos when the film was completed. *(Paramount, Charles Rector)*

"CATCH-22" REMAKE

In continuation of the previous text, it is interesting to note the following. In 2018, the Italian island of Sardinia was hosting a small fleet of warbirds, including two B-25J Mitchell bombers, a Douglas Dakota, and a Junkers Ju 52/3m. These airplanes were taking part in filming for a new adaptation of Joseph Heller's satirical "Catch-22". The production was set for release in 2019 as a six-episode, limited run series for the web-streaming television service HULU. It will also air on Channel 4 in the U.K. and Sky Italia in Italy, who are co-producing the series alongside HULU. The series, a co-production between the three networks, stars Christopher Abbott, Kyle Chandler, Hugh Laurie, and George Clooney, who is also set to executive produce alongside Grant Heslov.

The two B-25s are N3675G, s/n 44-30423, "Photo Fanny" of the Planes of Fame Air Museum at Chino, California and N898BW, s/n 45-8898, "Axis Nightmare" of the Tri-State Warbird Museum at Batavia, Ohio. The airplanes were based at Olbia-Costa Smeralda airport in north-eastern Sardinia, but occasionally they landed at Olbia-Venafiorita, a small airport south of the main airport. Most of the air-to-air scenes were filmed around the city of Nuoro. Both airplanes were painted in the Olive Drab/Neutral Grey USAAF colour scheme and provided with fictitious names.

STEVE HINTON

One of the modern time aerial photographers in the U.S.A. is Steve Hinton. He grew up around warbirds at the Planes of Fame Museum in Chino, California. Hinton learned how to fly for the camera from the masters, raced, crashed, and won gold at Reno. Steve Hinton was born on 1 April, 1952, in China Lake, California. When Hinton was seven years old, his parents bought a house in

"Photo Fanny" was transformed into the fictional "Fly Me High!". The B-25 is taxiing at Olbia-Costa Smeralda airport on Sardinia. One of the movie cameras is protruding from the tail gunner's position. The waist escape hatch is open too. "Axis Nightmare" was transformed to represent the fictional 6A "Yankee Doodle". (Fabio Ledda, Onnis Gian Luca Sardegna)

Claremont, California, next to Ed Maloney. Ed's son Jim became Hinton's best friend. They grew up together, became pilots, and worked together in the museum that Ed had started, the Planes of Fame Museum. Hinton soloed at 19 and has been immersed in warbirds ever since. Air shows and racing at Reno brought him in contact with a group of pilots who flew for the films and TV. He began his motion picture and television aviation career in 1976, flying a variety of World War II warbirds for the television series "Baa Baa Black Sheep", also known as the "Blacksheep Squadron". He also won races at Reno and Mojave. Hinton's work for films found him switching hats as a pilot in front of the camera, a pilot of the camera ship, aerial coordinator for the flight scenes, and preparing warbirds. He has worked for more than 60 films, including films with B-25s as "Forever Young" and "Pearl Harbor". For the aerial film work he uses the B-25 "Photo Fanny" from the Planes of Fame Air Museum. For the film "Pearl Harbor", Hinton put a gyro-stabilized 35mm camera in the tail of this B-25. Steve Hinton is a charter member of the Motion Picture Pilots Association and he is President of the Planes of Fame Air Museum at Chino.

The B-25 "Photo Fanny", described with the Planes of Fame Air Museum, has the civil registration N3675G and was obtained by The Air Museum of Ed Maloney in the mid-1960s. The Air Museum moved to Chino in 1973 and was renamed Planes of Fame Air Museum. The airplane has been kept in flying condition since the early 1960s. She regularly appears at air shows throughout the southwest and is frequently used for film and television projects, both as a camera platform and as a subject for various photo

Both nose art paintings, "Yankee Doodle" and a screen shot of "Fly me High". (Carl Roberts, Paramount TV)

Nice view of the filmset with the two B-25s. (Paramount TV)

An early picture of N3675G at the Air Museum, Ontario, California, 1970. (Chris Kennedy)

N3675G in the 1990s, she was overall natural aluminium finished and named "Photo Fanny". For some time, she had red/ white striped rudders and blue vertical stabilisers. (Collection Wim Nijenhuis)

Here some pictures of the airplane in 2008, with details of her specially moulded plastic hemispherical nose section. (Gary T. Takeuchi)

Nice pictures of "Photo Fanny" in flight with tail camera position manned.

(Collection Wim Nijenhuis)

Steve Hinton flew "Photo Fanny" from Chino for TV and films and used it as a camera ship. In 2000, she was used as a camera ship for the film "Pearl Harbor". Here, she is fitted with a Spacecam gyro-stabilized ball mount in the tail gunner's position. Spacecam offers a variety of solutions to adapt capabilities usually offered only by ground-based filming, to aerial cinematography. (Collection Wim Nijenhuis)

projects. So, the airplane has been both in front of and behind the cameras in many projects, films and TV-work. Over the years, the airplane has spotted different colours and names. She flew as "Shangri-La Lil" and later as "Betty Grable". The last name "Photo Fanny" was adopted during the filming of "Iron Eagle III". The nose art is based on that carried originally by a photo-recon B-24 Liberator and was painted by Teresa Stokes. The ship appears in several films like "Iron Eagle III", "Air Force One", "Forever Young" and flew in 2000 for the film "Pearl Harbor" as a camera ship and in the Hawaiian sequences as a Doolittle Raider with serial number 02261 in WWII USAAF camouflage. In "Pearl Harbor" she flew off the carrier USS Constellation. For filming "Pearl Harbor" a gyro-stabilized 35mm camera was put in the tail. It is the only fixed-wing plane ever to carry a gyro camera.

The airplane has been modified to serve as a camera ship. It has been wired with an extensive intercom system, and it has additional interphone cables throughout the airplane if additional positions are needed. A Plexiglas astrodome has been mounted where the top turret had been. This is useful for aerial directors to use during filming. It has also portable video screens. The greenhouse nose can be replaced with a specially moulded plastic hemispherical nose section when needed. The complete tail cone can be removed when needed to provide a direct view from the tail. There are also camera positions behind the glass of the waist gun blisters and the opening for the emergency exit hatch on the right side of the rear fuselage. The airplane is still in flying condition and is used as a camera ship and taking part in air shows.

B-25s WORLDWIDE

CANADA

The B-25 of Aurora Aviation in her colourful blue, white and grey paint scheme. The letters CF–DKU are applied in red on her vertical tails. Below the cockpit a small Canadian flag has been applied. (Collection Wim Nijenhuis)

The Royal Air Force squadrons in the 2nd Tactical Air Force flew B-25 Mitchells in Europe. Many Royal Canadian Air Force officers and men were attached to the RAF units that operated the type. Following the war, the Mitchell was then supplied in quantity to Royal Canadian Air Force Auxiliary Squadrons along with various other units. It was used primarily as a pilot, navigational or radar trainer and as a high-speed transport until its retirement in 1962. Many post-war RCAF Mitchells incorporated a new exhaust system where the top S-shaped stacks were replaced with semi-collector rings. Surplus Mitchells were often used as an aerial tanker and some examples were converted to other uses or sold. Hereafter, several examples have been given of companies that have flown with the B-25 as well as of aircraft displayed in Canadian museums.

AURORA AVIATION LTD., EDMONTON, ALBERTA

One of the civil B-25s in Canada was a B-25J-35 with s/n 45-8835. In 1972, she was sold to Aurora Aviation Ltd. of Edmonton, Alberta. Jack Rees and three other partners, all ex-RCAF crew members with some experience with the B-25, were involved in the company. The plan was to haul diesel fuel to isolated areas in the Canadian Arctic. So, in the spring of 1972, they started hauling diesel fuel from Yellowknife, North-west Territories to Hope Bay silver mine approximately two hours flying time north on Bathurst Inlet. They were carrying approximately 1,300 gallons and landing on ocean ice. They studied up on the practicalities of fuel hauling and decided that a B-25 would make a suitable and economic tanker. They located the B-25J s/n N5672V in New Jersey, one of several used over the years in flight research by Bendix Aviation. In September

1972, the airplane was sold to Aurora for $ 20,000. She was ferried to Edmonton, registered as CF-DKU and was converted into a tanker. Several tanks were installed: one in the nose, dorsal turret, bomb bay and rear fuselage. Tankage totalled 1,400 Imp. Gallons. She was painted in a colourful blue, white and grey paint scheme. In 1973, they started hauling diesel fuel from Smithers, British Columbia to Chipmunk Creek for a railway right of way extension. During her fuel-hauling years, she often flew two trips per day. The B-25 handled this service for some time until one day a nose wheel blew out while landing loaded in Chipmunk Creek. The blow out caused the nose gear to collapse resulting in the propellers eating dirt. No one was hurt. After the plane was repaired and put back into service, she flew for another year hauling fuel. But once again in 1974, another nose wheel accident happened. The insurance company refused to pay this time and bought CF-DKU from Aurora Aviation and put her up for auction. In 1975, G&M Aircraft of St. Albert, Alberta won the bid and owned her right up to 1991 flying the B-25 as a water bomber.

AVACO SERVICES, SALMON ARM, BRITISH COLUMBIA

Little is known about four Mitchells registered in Canada as CF-NTU, CF-NTV, CF-NTW and CF-NTX. They were all purchased by Avaco Services of Salmon Arm, British Columbia in 1962 after they were sold by the Canadian Government. All four were ferried to Kamloops, British Columbia for "conversion" which never happened, and then were left derelict. Kamloops Airport, also known as Fulton Field or Davie Fulton Airport, is a regional airport located northwest of Kamloops. Only CF-NTU made it out intact and became a gate guard at CFB Winnipeg after 1974, where she remains to this day.

CF-NTU

This was an ex-USAF B-25J-30 with s/n 44-86724. She was noted "for disposal" on 5 January, 1961 and pending disposal at RCAF Station Lincoln Park, Alberta from 17 March, 1961. Then sold to Avaco Services and registered as CF-NTU. She was ferried to Kamloops Airport in 1962. She was stored outside, derelict, until at least 1971. In 1974, she went to the Canadian Forces for restoration and was displayed at CFB Winnipeg by 2002.

CF-NTV

This B-25 was an ex-USAF B-25J-30 with s/n 44-31399. She went to inactive reserve with RCAF Station Lincoln Park on 7 October, 1960, and was stored. She was sold to Avaco Services and was stored outdoors at Kamloops Airport for many years, seen there as late as 1974.

CF-NTW

A third B-25 was s/n 44-86728. This was also an ex-RCAF machine stored and sold to Avaco Services and registered as CF-NTW. She was also reported derelict at Kamloops in the 1970s.

CF-NTX

The fourth plane was a B-25J-30 with s/n 44-31493. She was sold to Avaco Services and registered as CF-NTX. She was reported derelict at Kamloops airport in 1974.

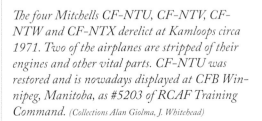

The four Mitchells CF-NTU, CF-NTV, CF-NTW and CF-NTX derelict at Kamloops circa 1971. Two of the airplanes are stripped of their engines and other vital parts. CF-NTU was restored and is nowadays displayed at CFB Winnipeg, Manitoba, as #5203 of RCAF Training Command. (Collections Alan Giolma, J. Whitehead)

CF-NTX in the weeds off the runway at Claresholm, Alberta in July 1970. (Blake Smith)

G&M Aircraft Ltd., St. Albert, Alberta

G&M Aircraft Ltd. was a firefighting company based at Fort Smith, a town located on the southwest bank of the Slave River near the Alberta-Northwest Territories border, north of Edmonton. G&M Aircraft had a fleet of three B-25s fighting fires in Wood Buffalo National Park. Wood Buffalo is Canada's largest national park. It sprawls across North-Eastern Alberta and juts into the southern part of the Northwest Territories. Under chief pilot Jack Rees, former pilot with Aurora Aviation Ltd. of Edmonton, G&M held the firefighting contract for the park until 1991, when Ottawa decided no longer to fight fires in Wood Buffalo. So, G&M Aircraft used the Mitchells until 1992.

In May 1975, C-FDKU was converted into a water bomber with a new nose and new overall blue colour for G&M Aircraft at Edmonton. In a later stage, the paint on the underside had been removed. On the rear fuselage in small letters is the company name G&M Aircraft.

(Collection Wim Nijenhuis)

C-FDKU

In 1975, G&M bought C-FDKU from Aurora Aviation. She was registered as C-FDKU and flew as tanker number 1, later with number 336. She was converted into a water bomber at St. Albert and was part of the G&M fleet. The airplane was painted in an overall blue colour scheme with yellow striping. Later, the paint on the underside was removed and at the time she had number 336, she was painted in an overall yellow paint scheme. In 1993, she was sold back into the U.S., where she became N5672V. After several years being restored, she flew again in 1999. Today she is warbird "Betty's Dream" in Texas.

Again C-FDKU but now painted in an overall yellow paint scheme with number 336.

(Tom Tessier)

C-FMWC

In 1981, a second B-25 was purchased. The B-25J-25 s/n 44-30254 was sold to G&M Aircraft from Northwestern Air Lease. This ex-RCAF #5211 was registered as C-FMWC. The company operated her as tanker #2 and later as tanker #337. In June 1995, she was sold to Vintage Wings from Anchorage, Alaska.

C-GUNO

The third B-25 of G&M Aircraft was s/n 44-86698, a B-25J-30. She was in service with G&M from August 1982 until 1997. In May 1983, she was registered as C-GUNO and flew in an overall yellow paint scheme as fire bomber #3, later as #338. She was withdrawn from service in 1992. She was the last operating B-25 fire bomber.

C-FDKU during a flight in 1993.

(Richard de Boer)

The second tanker of G&M with number 337 was this B-25 registered as C-FM-WC. She was also painted yellow. (Neil Smith, Geoff McDonell)

The third overall yellow painted B-25 of G&M Aircraft. The company name is written in small letters on the rear fuselage and this ship was numbered 338. The Canadian registration C-GUNO is applied on the rudder. This was the last B-25 in the world to be operated as a fire bomber. She was withdrawn from service in 1992. (napoleon130.tripod.com)

Hicks and Lawrence Ltd., Ostrander, Ontario

Hicks & Lawrence Limited owned an ex-RCAF B-25J-30, s/n 44-86698. She was sold to this company in December 1961 and then registered as CF-NWU. In May 1965, the company sold her again to Eldon Armstrong of Toronto. Another B-25 of the company was a B-25D-35, s/n 43-3634. This airplane was delivered to the RCAF as KL148 in October 1944. She was struck off charge in June 1962 and sold to Hicks & Lawrence. A provisional registration CF-NWV was assigned for the ferry flight to St. Thomas in August 1962. She was stored outdoors there until 1969, when she was sold to Richard McPherson/Aerodix, New Albany, Ohio, and registered as N3774.

Hicks & Lawrence is an air operator, providing air charter services and airspace and aircraft management services. In 1947, the company was founded by Mervin A. L. Hicks and Tom Lawrence. Hicks instructed pilots from schools near St. Thomas and London.

Lawrence Mitchell and Tom Lawrence were in Hicks' first class of student pilots. Lawrence was to be in charge of the agricultural end of the business. Hicks operated the Tillsonburg Airport, Ostrander, Ontario, from 1946 to 1962, and during those years the airport became known as the Hicks and Lawrence Airport. Later, he also operated the St. Thomas Airport. Hicks and Lawrence Ltd. was given approval in June 1965, to operate a commercial helicopter service from bases in Tillsonburg and St. Thomas. The company was first marketed agriculturally in the large local tobacco market. Based at Tillsonburg Airport, they purchased war surplus aircraft, including many Harvard's and B-25s. Mervin's son Duane Sr. expanded to St. Thomas and operated the airport there, while expanding into the North including fire bombing and birddog operations in Dryden. They also operated Grumman TBM Avenger aircraft equipped with a hopper and spray nozzles, spraying in New Brunswick every year from 1969 to 1985. Duane Sr. took over the company in 1987 and a year later obtained a large fire detection contract with the Ontario Ministry of Natural Resources. At this point, Hicks & Lawrence was operating up to 38 aircraft

Hicks & Lawrence Ltd. owned an ex-RCAF B-25J-30 registered as CF-NWU. The airplane is still painted in the RCAF colours. (Brian Stainer via Bob Hobbs)

between spray, detection, and birddog operations. Duane Jr. took over aerial (spray) applications in 1989 and sold the spray business less than 8 years later to focus on aerial firefighting. He purchased the rest of the company in 2001. Nowadays, the company is a subsidiary of Discovery Air Inc.

The other B-25 was sold to Hicks & Lawrence in 1962. This B-25 was delivered to the RCAF as KL148. In 1954, she was at the Flying Instructors School at RCAF Station, Trenton, Ontario. There, she had a landing accident that damaged the greenhouse nose, and a solid nose was installed. She was struck off charge in June 1962 and sold to Hicks & Lawrence. A provisional registration CF-NWV was assigned for the ferry flight. In 1969, she was sold to Richard McPherson/Aerodix, New Albany, Ohio, and registered as N3774. (b-25history.org, Bill Stainer via Bob Hobbs)

Holden Aviation Services, Ltd., Lamont, Alberta

C-GTTM in an overall natural aluminium finish. She still has the nose art of her former U.S. owner Earl "Red" Dodge, Anchorage, Alaska, where she flew as tanker #4. This was a wonderful nose art depicting a horse pulling an old-fashioned fire wagon. (Collection Greg Rees)

In 1980-1981, for a short time the B-25J-5 with s/n 43-28059 was property of Holden Aviation Services at Lamont, Alberta. The airplane was modified by Hayes in 1954, and by 1957 she was flown to storage. In 1959, she was fitted with a 1,000-gallon tank. After she had a few American owners, she was sold to Holden Aviation Services in September 1980 and registered as C-GTTM. But already in May 1981, she was purchased by Aero Trader in Borrego Springs, California. She underwent an extensive restoration and was sold to Kermit Weeks in May of 1983. She now resides at the "Fantasy of Flight" museum in Polk City, Florida.

Northwestern Air Lease Ltd., St. Albert, Alberta

Northwestern Air Lease Ltd. is based in Fort Smith. The family-run airline grew from a small-scale water bombing operation to transporting thousands of people in the South Slave Region. Terry Harrold, the airline's founder, started flying in the 1950s, training with the air cadets and Royal Canadian Air Force.

In 1964 he started flying with Air Spray as a water bombing pilot. One year later in 1965, he and his partner Peter Kuryluk formed Northwestern Air Lease and started getting work "bombing" forest fires with old Second World War planes. Their first aircraft was a Beech 18. Harrold and his partner started converting a Second World War Lancaster

bomber to fight forest fires in Alberta. In 1971, Peter was killed when flying a Canso water bomber fighting a fire. Shortly after, Terry and his wife Ruth bought his shares and therefore wholly owned Northwestern. Their son, Brian, became a full partner in Northwestern in 1981. They showed up at different places where they knew the forest fires were and tried to get the work. Nowadays, the company has over 70 employees and is the main airline operating out of Fort Smith providing daily scheduled flights and chartered services for Alberta and the Northwest Territories.

Northwestern Air Lease operated two B-25s as water bombers. They had the Canadian registration CF-OND and CF-MWC. The home base for these Mitchells as previously mentioned was St.Albert Airport, which consisted of a dirt runway at the time and was more like a mud runway after it rained. In 1974, both B-25s flew in Wood Buffalo National Park. 1975 was just as much a mystery as the year before when it came to operations. During the early years of water bombing, everyone involved shared the

Tanker CF-MWC operated by Northwestern Air Lease is refuelling at Watson Lake in 1978. (John Poirier)

A close-up of both Northwestern airplanes. At left, CF-OND and at right CF-MWC. Note the different noses. CF-MWC still has the air force radome nose and is equipped with a shark nose motif. (Collection Wim Nijenhuis)

excitement and enthusiasm when called out on a fire. So even when the fires were small and insignificant, the B-25 crews were hipped up and raring to go if it meant some flying time.

The B-25 s/n 44-28866, a B-25J-15, was registered as CF-OND and was sold to North Western Air Lease in 1970. She flew as tanker number 90. By October 1993, she was sold to BHA Leasing of Carmel, Main. In 1966, the B-25J-25 s/n 44-30254 and regis-

tered as CF-MWC was sold to Northwest Air Lease and used as a tanker. She was again sold in 1981 to G&M Aircraft from St. Albert.

For a very short time, Northwestern Air Lease had a third B-25. This was a B-25J-25 with s/n 44-30456 and was registered as C-GTTS. In 1982, this B-25 was bought from Air Spray Ltd., Edmonton, Alberta, for a water bomber role. But the B-25s were phased out of the water bomber role, and this B-25 was soon sold to William G. Arnot of Breckenridge, Texas and flown back to the USA and was registered as N43BA.

Years later, CF-OND in an overall white fuselage with a bright red lightning flash. The engines nacelles are black with a yellow cowling ring. Note the black and white striped propeller blades. In this beautiful and atmospheric picture, the airplane is seen in March 1989, between the fire seasons at Hay River, Northwest Territories, Canada. (Gary Vincent)

ROBERT DIEMERT, CARMAN, MANITOBA

On 28 June, 1964, a new, 3,500-foot-long grass airstrip, Friendship Field, was officially opened at Carman, 40 miles southwest of Winnipeg. Although the field was privately owned by Robert Diemert, it would be used as a home base for what appears to be the first co-operative Flying Club in North America. During the 1970-1980s, from this private Friendship Field Airport and restoration shop at Carman, Manitoba, Diemert was involved with the restoration of several warbirds. Diemert is one of the first warbird restorers in the world. He restored a Hawker Hurricane for use in the movie The Battle of Britain and flew as a stunt pilot in that film. With the funds made from that project, he travelled to the South Pacific in 1968-1969 to look for aircraft to recover. Robert Diemert has restored a number of historic aircraft for clients in the USA, including Japanese fighters.

Diemert owned one B-25. This B-25 was purchased by him in November 1972 from A.C. Ellis, Galveston, Texas. This was a B-25J-25 with s/n 44-30456 and was registered in the civil register as N3512G. In June 1967, this B-25 was modified with a waist cargo door on the left side just aft of the wing and was used to transport cargo. She had been used extensively for hauling freight across the Caribbean. This B-25 was purchased by Bob Diemert in 1972 in Texas. The aircraft was to be scrapped and Diemert bought both engines. The owner did not want the airframe, so Diemert left the engines on the aircraft, and flew it in 1973 to Carman, Manitoba and registered her as C-GTTS. In 1981, Diemert sold the B-25 to Air Spray Ltd., Edmonton, Alberta, and in 1982, the airplane was sold for a water bomber role to Northwestern Air Lease, Edmonton. But the B-25s were phased out of the water bomber role, and this B-25 was soon sold to William G. Arnot of Breckenridge, Texas and flown back to the USA.

A colourful C-GTTS photographed at Carman in the 1970s, but still with the U.S. registration N3512G. Note the dismantled rudders. (Dave Welch, Collection Wim Nijenhuis)

ALBERTA AVIATION MUSEUM, EDMONTON, ALBERTA

The Alberta Aviation Museum is a museum in Edmonton, Alberta, and has one B-25. The museum's bomber is a gift from Terry and Brian Harrold. This B-25, s/n 44-30791, was originally taken on strength by the RCAF in August 1953, spending most of its service life at No. 2 Air Navigation School in Winnipeg before being retired in February 1962. She had the RCAF number 5273. In 1962, she was sold to Northwestern Air Lease at St. Albert and was stripped for parts for their B-25 tankers. She never became a civil Canadian registration number. The hulk of the airplane was stored at the farm of Terry Harrold, owner of Northwestern Air Lease. The Harrolds of Northwestern Air Lease were active in purchasing several B-25s after the war, most of which were converted into air

For years the B-25 of the Alberta Aviation Museum sat derelict at the farm of Terry Harrold near St. Albert. Although the fuselage was largely intact, the wings, engines and propellers had been taken as spare parts to keep a fleet of fire bombers flying. This picture was taken in 2000. (Tanker336)

After restoration, the nose art on the left side recognizes the success of the Edmonton Eskimos in their Grey Cup wins of the 1950s. (Collection Wim Nijenhuis)

tankers for forest firefighting purposes. In April 2002, the hulk was trucked from the farm to the Alberta Aviation Museum. When donated by the Harrold family, the airplane had been derelict at their family farm for many decades having been part of their water bomber operations. The airframe was missing many components and had suffered from some damage. She was stored dismantled in original RCAF scheme marked as #5273 and code SV-273. From 2005 until 2011, she was restored for static display and rolled out for display on 3 September, 2011. She is displayed in a RCAF scheme as FW251 code HO-251 and named "Daisy Mae". When the airplane was restored, it was decided to commemorate one of No. 418 Squadron's Mitchells from the 1950s, HO-251, which was involved in the infamous crash into the side of the hangar that today houses the Alberta Aviation Museum. The nose art on the left side recognizes the success of the Edmonton Eskimos in their Grey Cup wins of the 1950s. It replicates the art used on some 418 Squadron aircraft during the 1950s when the squadron flew the Mitchells. On the right side, the airplane carries the character "Daisy Mae" from Al Capp's "L'il Abner" comic strip as nose art. Today, she is a static display at the museum, housed in Canada's last double-wide double-long wartime hangar of the British Commonwealth Air Training Plan. These hangars, built for the BCATP across Canada, were made of precut wooden timbers of British Columbia fir. They could be built as single units, double units, and the 'double-double' which is four units. The Alberta Aviation Museum opened in 1992 as a non-profit historical society dedicated to preserving Alberta's aviation heritage. It was founded by a large group of aviation enthusiasts, many of whom worked or trained in the aircraft hangar that houses the museum.

On the right side, the airplane carries the character "Daisey Mae", a character in Al Capp's comic strip "Li'l Abner". This was a satirical American comic strip that appeared in many newspapers in the United States, Canada, and Europe. It was written and drawn by Al Capp (1909–1979). The strip ran for 43 years, from 1934 through 1977. (Alberta Aviation Museum, Lightjug)

CANADIAN WARPLANE HERITAGE MUSEUM, HAMILTON, ONTARIO

The Canadian Warplane Heritage Museum is a Canadian aviation museum located at the John C. Munro Hamilton International Airport in Hamilton, Ontario. The museum was founded in 1972 and is a non-profit organisation whose mandate is to acquire, document, preserve and maintain a complete collection of aircraft that were flown by Canadians and the Canadian military from the beginning of World War II to the present. The founders Dennis Bradley, Alan Ness, Peter Matthews, and John Weir became partners and acquired the first

In late 1975, C–GCWM was purchased by the Canadian Warplane Heritage Museum. Museum President Dennis Bradley congratulates pilot Harry Kelly on the delivery of the museum's B-25. For the flight she was nicknamed "She's a Silver Lady". At the time, the airplane was overall natural aluminium finished.

(Canadian Warplane Heritage Museum)

airplane, a Fairey Firefly. This aircraft was to become the masthead of the Museum's advertising and stationery and continues to this day to be incorporated into logos, crests, and memorabilia. In 1972, the group moved into part of a hangar at Hamilton Airport and started to seriously seek out other restoration projects or flying aircraft. A de Havilland Canada Chipmunk was the second acquisition, followed shortly by the Supermarine Seafire, Corsair, Harvard and Tiger Moth. The group applied for foundation status, to be governed by its own volunteers, operating as the Canadian Warplane Heritage. 1975 saw the collection move into another area in Hangar #4 and the acquisition and restoration of the B-25 Mitchell began. The museum now houses over forty aircraft, and an extensive aviation gift shop and exhibit gallery.

C-GCWM

The museum's B-25 is a B-25J-35, s/n 45-8883, that never saw military service and was operated as a civilian transport for over 25 years. Her registration was assigned as

N75755 in January 1947. After she had several U.S. owners, she was purchased by the Canadian Warplane Heritage Museum. In 1975, the museum found her abandoned at Wilmington Airport, Delaware. After repairs, she was marked as "She's a Silver Lady" and flown to the museum, where she underwent extensive restoration. During the fall of 1975, the B-25 was moved into the hangar and the restoration of the aircraft began. New engines were installed, all fuel, oil and hydraulic systems were overhauled, and the civilian interior was replaced with a military one. Her registration was reserved as C-GCWA. This was later changed to her current registration C-GCWM. Her first flight was on 8 May, 1976, and she was painted in the colour scheme of HD-372 of No. 98 Squadron RAF as flown by RCAF F/O J. W. David Pudney of Vancouver, BC. In 1992, she got her name "Hot Gen!". For a short time around 2005, she was named "Grumpy". Her original nose art has been recreated and currently she is flying again as "Hot Gen!" The original "Hot Gen" was B-25D-35 of the RAF with serial FW-275 and USAAF s/n 43-3720. She survived the war and was struck off charge on 27 September 1945. The Canadian pilot of "Hot Gen" was F/L A.L. Duncan. George Van Iderstine was Air Gunner on that crew, and, in later years, Van Iderstine was a member of the Canadian Warplane Heritage that renamed their Mitchell "Hot Gen" in his honour.

C-GCWJ

In June 1980, a B-25 was reregistered as C-GCWJ in the Canadian civil register to the Canadian Warplane Heritage at Sidney, British Columbia. This was a B-25J-35 with s/n 45-8884. In 1978, Jerry C. Janes of Vancouver acquired this B-25 which had been op-

In 1976, the airplane was painted in the colour scheme of HD-372 coded VO-D of No. 98 Squadron RAF. In 1992, her aircraft letters were changed to VO-F and named "Hot Gen!". However, for a short time, she was repainted as VO-B "Grumpy". At the time, she had the black/white invasion stripes. (Canadian Warplane Heritage Museum)

In 2011, the original nose art had been recreated. Artist Lance Russwurm was asked to paint the "Hot Gen" nose art on the fully restored B-25 bomber. Note the difference in the nose art from 1992 (left) and the latest of Lance Russwurm. (Collection Wim Nijenhuis)

erated by the Minnesota Aircraft Museum. The airplane was registered in the U.S. as N3156G. While she still was operated by Jerry C. Janes, the airplane flew for a short time as 458884 nicknamed "The Death Watch", but later received the markings #5262 and HQ-262. In 1980, the airplane was flown to Edmonton where she was restored by members of No. 418 (Auxiliary) Squadron. B-25 Mitchells were predominately operated by the post-war RCAF with No. 418 "City of Edmonton" Squadron and No. 406 "City of Saskatoon" Squadron from 1946 until 1958 assigned to the light transport and emergency rescue roles. A number of Mitchells were also operated by the Air Navigation School as navigational trainers. The restored plane was finished in the colours and markings representing a B-25 operated by No. 418 Squadron in the mid-1950s, "City of Edmonton", RCAF #5262 and code HQ-262. On the right side, the airplane

had a nose art of Daisey Mae, a character in Al Capp's comic strip Li'l Abner. The airplane was operated by Jerry Janes on behalf of the Canadian Warplane Heritage at Sidney, British Columbia. This was a west coast

"The Death Watch" in 1979. The ship, s/n 45-8884, was overall natural aluminium finished with a nice nose art on the right side and black outer surfaces of the vertical tails. (Collection Wim Nijenhuis)

By 1982, "The Death Watch" was finished in the colours and markings representing a B-25 operated by No. 418 Squadron in the mid-1950s. She was named "City of Edmonton" and had number 5262 and code HQ-262. On the left side of the nose the name "City of Edmonton" and the squadron emblem were applied. *(Den Pescoe)*

In August 1985, C-GCWJ returned to the U.S. and registered as N5833B and flew as "Georgia Girl" with Randal Porter and David Brady at Cartersville, Georgia. Both names are written below the cockpit. Here she is seen in March 1990 at the Valiant Air Command Airshow, Titusville, Florida. *(Robert Bourlier)*

branch of the Canadian Warplane Heritage that never quite took off. The B-25 C-GCWJ graced the skies for a few years in the attractive post-war colours of the RCAF. In August 1985, the aircraft returned to the US register as N5833B, registered to Mark Clark of Courtesy Aircraft, Rockford, Illinois. The same month, the aircraft was reregistered to Randal Porter and David Brady of Air Acres Air Museum, Cartersville, Georgia. The aircraft was named "Georgia Girl". Chris & Patrick Harker of C&P Aviation Services, Anoka, Minnesota, became the next owners in August 1993. In September 2004 the aircraft was transferred to Chino, California, where the aircraft underwent a long restoration to military standards and on 15 May, 2007, she was delivered to her present owner, Lady Luck LLC. of Blaine, Minnesota. The aircraft was renamed "Lady Luck".

REYNOLDS-ALBERTA MUSEUM, WETASKIWIN, ALBERTA

A B-25 was displayed at the Reynolds-Alberta Museum in Alberta. This museum in Wetaskiwin, Alberta, interprets the impact of technological change in transportation, aviation, agriculture, and industry from the 1890s to the present. The museum opened in 1992 and is a project of Alberta Culture and Tourism and Alberta Infrastructure. The museum is named after Mr. Stan Reynolds, a Wetaskiwin businessman and world-renowned collector who donated a core collection of 1,500 artefacts to the Province of Alberta. Mr. Reynolds donated this core collection between 1982 and 1986, but also continued to donate artefacts regularly from his collections until his death in February of 2012.

From 1992, the museum displayed an unrestored B-25J-30 with s/n 44-86726. This was an ex-RCAF Mitchell numbered 5237. There is some misunderstanding about this plane. Many sources indicate that this is CF-NTP. However, as far as is known, Canadian registers indicate that #44-86726 has never received a civilian U.S. or Canadian registration. Unfortunately, the former RCAF Mitchell was stored outside the Wetaskiwin airport and weathered considerable and has seen a lot of vandalism. The remains of the RCAF are still somewhat visible.

N5833B of Lady Luck LLC. of Blaine, Minnesota and flown by Patrick Harker. *(D. Miller)*

This is the Canadian B-25 in her better days, an ex-USAF B-25J-30/32, s/n 44-86726. She was received from the USAF in February 1952 and was struck off charge in October 1960. She had the RCAF number 5237 and aircraft code QP-237. At right, the same airplane in 2014 and unfortunately in derelict condition in a yard next to the Reynolds-Alberta Museum in Wetaskiwin. *(Collection Wim Nijenhuis, John Olafson)*

OTHER CIVIL CANADIAN B-25S

CF-NTP

An airplane that had only for a very short time a civil Canadian registration was a B-25J-30 with s/n 44-30947. In October 1951, she was taken on strength by Air Defence Command and numbered 5212. After her service with the RCAF, she went to storage at RCAF Station Lincoln Park, Alberta in December 1960. She was pending disposal there from February 1961. She was registered as CF-NTP for her ferry from Lincoln Park to Calgary, Alberta, in October 1961. In 1962, she was sold to Columbus L. Woods/ Woods Body Shop, Lewistown, Montana, and registered as N92880. Nowadays, her forward fuselage section is being used to rebuild the "Sandbar Mitchell" at the Warbirds of Glory Museum at Brighton, Michigan.

CF-NTS

The Fairey Aviation Company Limited of Canada owned a B-25J-25 with s/n 44-30641. This was an ex-RCAF ship with number 5257. After her RCAF service, she went to inactive reserve at RCAF Station Lincoln Park, Alberta on 9 December, 1960 and pending disposal there from 27 February, 1961. She was sold to Fairey Aviation of Canada, Victoria International Airport and registered as CF-NTS.

The Fairey Aviation Company Limited was a British aircraft manufacturer founded in 1915 in Hayes, Middlesex. The factory had a strong presence in the supply of naval aircraft, and also built bombers for the RAF. Formed in 1948, the Fairey Aviation Company of Canada grew from a six-man operation to a major enterprise employing around a thousand people. From 1949, the company undertook repair and overhaul work for the Royal Canadian Navy. Later it also undertook modification work, conversion programmes and overhaul for the RCAF. The West Coast Branch of the company was formed in 1955 at Sidney. The plant was located at Victoria International (Patricia Bay) Airport. This facility handled mainly repair, overhaul and modification of military and civil aircraft including the conversion of ex-military Avenger aircraft to commercial crop-dusting roles.

CF-OGQ

For a number of years, a B-25 was owned by Joe E. Goldney, Vancouver, British Columbia. This was a B-25D-30 with s/n 43-3318. The RCAF used her as a target tug when all other duties ran out and the machine served at both Cold Lake and Uplands in this role before being placed into storage at Claresholm in 1960. In February 1962, she was struck off charge and she was acquired from the Crown Assets Disposal Corporation by Joe E. Goldney from Vancouver and registered as CF-OGQ. She was flown to Vancouver the same month. She would stay in Canada until she was bought by her first U.S. owner in March 1966, being Sports Air Inc. in Seattle, Washington and registered as N88972. Nowadays, the airplane is a famous warbird of the Historic Flight Foundation at Paine Field, Washington, and she is flying in RAF colours as "Grumpy".

CF-OVN

The ex-RCAF B-25 #5272 was a B-25J-25 with the U.S. s/n 44-30421. On 7 October, 1960, she went to inactive reserve with RCAF Station Lincoln Park, Alberta. She was struck off charge in February 1962 and stored at Claresholm, Alberta. She was sold to Canspec Air Transport Ltd. This company, headquartered in Calgary, Alberta, offered specialised air services, including firefighting, spraying, photo survey, long range patrol and magnetometer survey. The company was founded by T.R. Reynolds in 1961 and had also five PBY 5A Canso twin-engine amphibious aircraft for their operations. Later, the B-25 was sold to I.J. Dowler of Vancouver, British Columbia. She was civil registered as CF-OVN. But probably she never flew as CF-OVN and was stored at Vancouver with "Horizons Unlimited - Canada" logo until she was sold to Robert Sturges of Troutdale, Oregon in January 1969 and registered as N7674. During WWII Robert Sturges had been a Boeing technical representative in Britain on B-17s and after the war set up Columbia Airmotive Inc. as an aircraft and parts dealership. N7674 was flown as warbird "Dirty Gertie from Bizerte" for some time. She was sold in June 1978 to the Historical Aircraft Preservation Group at Borrego Springs, and eventually ended up in Colombia circa 1979. She is reported crashed around 1980 after the right main gear collapsed while landing on a dirt strip at Santa Marta, Colombia, during a drug run. The wreck was bulldozed to make way for other aircraft.

An undated picture of N88972, former CF-OGQ, possibly taken at Vancouver. She is still painted in her former RCAF colours. (Collection Tony Clarke)

CF-OVN at Vancouver in 1967, stored with "Horizons Unlimited - Canada" logo. She still has the RCAF number 5272 on her vertical stabiliser and the red nose cone. (Ken Fielding

UNITED KINGDOM

After the war, several civil B-25s have flown in the United Kingdom. They were all used as a camera platform for films and ended mostly in museums. In the decades after the Second World War, film makers had no computer-generated imagery (CGI). This is the application of the field of computer graphics or, more specifically, 3D computer graphics, to special effects in art, video games, films, television programmes, commercials, simulators, and simulation generally, and printed media. So, film makers had to use real cars, airplanes, boats, stuntmen, animals, etcetera. They had to film them with or without movement. For aerial shootings, the film makers needed real airplanes as an aerial camera platform. In that period, the airplane and the film camera have combined to produce countless motion pictures. From 1945 until about the 1990s, many films which had required an aerial camera platform, numerous television productions and various commercials have been filmed and photographed from B-25 Mitchell bombers. In the U.S.A. as well as in Europe, the B-25s used as a camera platform were mostly J-models. The B-25J was a particularly good airplane as a camera platform and was suited for a variety of film work. It offered an excellent visibility from the nose, the tail, and the left and right waist windows. In the United Kingdom, the Englishman John Hawke was the man behind several film productions on this continent. He also flew in some well-known films in a modified B-25 camera ship.

JOHN HAWKE

On the European continent, one man was leading in this business. That was John "Jeff" Hawke. John Hawke has been involved in films for many years along with Frank Tallman in the U.S.A. He has worked for many films involving the B-25. Hawke was a former pilot of the Royal Air Force. He was very capable in locating and flying film airplanes. He owned the companies Visionair International Inc. in Florida and London, Euramericair Inc. and Airspeed International

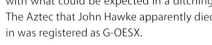

1978 in The Netherlands, John Hawke (centre) in front of one of the "Hanover Street" B–25s. John "Jeff" Hawke has worked for many films involving the B–25 in the United Kingdom. (Collection Wim Nijenhuis)

Sales Inc., both at Fort Lauderdale, Florida. These companies no longer exist. One-time President of Euramericair, he flew one of the Mosquitos in the film "633 Squadron." He also operated the B-25 camera ship for the film "Battle of Britain". As well as piloting the B-25 camera ship he also worked for "Empire of the Sun", "Sky Bandits", "White Nights", "Sweet Dreams", and "Hanover Street". Apart from "Hanover Street", he was aerial co-ordinator or advisor for all these films. In 1965, Hawke was hired to try to deliver twenty B-26 Invaders to Portugal for use in its African colonial wars. This, despite a U.N. embargo against arms sales to Portugal. John Hawke was also the man behind the five B-25s used in the film "Hanover Street" from 1979. In 1991, press reports said that John Hawke had hired a Piper Aztec. The airplane was located later in the Adriatic and recovered on 28 December, 1991. It could well have been lying there for up to 2 months. The body on board carried a Miami driving licence in the name of John Hawke. It was said that the undercarriage and flaps were down and showed damage inconsistent with what could be expected in a ditching. The Aztec that John Hawke apparently died in was registered as G-OESX.

N6578D

One of the B-25s of John Hawke was B-25J-30, s/n 44-31508, delivered to the US-AAF in June 1945. She was modified by Hughes in 1956 and assigned to Westchester AFB, New York and in 1958 stored at Olmstead, Pennsylvania. In 1960, she was registered as N6578D. In September 1967, after several changes of owners, she finally was sold to Euramericair Inc. of John Hawke. The airplane was to capture the spectacular aerial combat scenes for the film "Battle of Britain". This World War II film depicts the events surrounding the air defence of the British Isles by Royal Air Force pilots against the German Luftwaffe in 1940. The film was directed by Guy Hamilton and featured an all-star cast including Michael Caine, Christopher Plummer, Edward Fox and Robert Shaw. The film also took the extra step of hiring actual Germans to play Germans, speaking German. The film was released in

John Hawke taxis the camera ship in 1968. The airplane was known as the "Psychedelic Monster" due to its bright fluorescent paint scheme meant to make it easily distinguishable from the other airplanes during the filming of the aerial scenes. (Collection Adrian Allen)

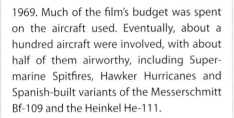

The "Psychedelic Monster" with the hemispherical optical bubble nose. The airplane was heavily modified to capture the spectacular aerial combat scenes for the film "Battle of Britain". A specially built clear vision camera position was installed in the tail, with slipstream deflecting shields above and below the camera position. It was awful windy for the cameraman. *(Collection Adrian Allen)*

1969. Much of the film's budget was spent on the aircraft used. Eventually, about a hundred aircraft were involved, with about half of them airworthy, including Supermarine Spitfires, Hawker Hurricanes and Spanish-built variants of the Messerschmitt Bf-109 and the Heinkel He-111.

A special camera platform was needed for the film as the primary aerial camera platform for the aviation sequences. The airplane was heavily modified in Florida by Hill Air and Flying W. Products. The B-25 was fitted with a hemispherical optical bubble which replaced the nose glazing, enabling a Panavision camera to shoot through 210o without distortion. Clear vision panels replaced the waist gun positions, the tail turret was removed, and a specially built clear vision camera position was installed with slipstream deflecting shields above and below the camera position. The only human comfort afforded the cameraman was a full harness seat belt. In the bomb bay there was a retractable double-jointed arm with a remote-controlled camera at the end which was capable of filming through 360o. In the mid-upper turret position was an enlarged astrodome below which the aerial unit director sat during filming, positioned around him was a bank of television monitor screens connected up to the cameras along with a video tape machine to enable instant playback of any particular camera shot. In December 1967, it took Hawke and co-pilot Duane Egli 22 hours to fly the B-25 from Florida to England.

On arrival the airplane was given a distinctive paint scheme. The forward fuselage was natural metal, leading edges of the wings were white, and the trailing edges of the wings were adorned with black and yellow stripes. The left rear fuselage was dayglo red and the right rear fuselage was green. This garish colour scheme had a practical purpose, being used to position different formations of airplanes in a particular sector of the B-25 camera ship. The markings were primarily intended for line-up references for aerial filming, and to make it easier for other pilots to determine which way the bomber was manoeuvring. Upon arrival at the filming location Tablada Air Base in Spain in March 1968, the spontaneous comment from Derek Cracknell, the first assistant director, was "It's a bloody great psychedelic monster!". The airplane was henceforth dubbed the 'Psychedelic Monster'. In January 1969, after completion of the filming, the airplane returned to the U.S.A. She was reported as derelict at Caldwell-Wright Field, New Jersey in 1970. In February 1979, after restoration by Tom Reilly, Kissimmee, Florida, she became airworthy again and flew as "Chapter XI" and later in the 1990s as "Lucky Lady". In 1998, based at Franklin, Virginia, she was withdrawn from use and placed in open storage. Eventually she degraded to derelict condition. In April of 2015, she was purchased by Reevers Warbird Roundup in Australia, restored, and painted as "Pulk", which was originally a B-25C of No. 18 NEIAAF Squadron.

N7614C

Another B-25 of John Hawke was s/n 44-31171, a B-25J-30. After storage at Davis Monthan, the airplane was sold to Radio Corporation of America, New York in 1958 and modified for use as a flying test laboratory. She was registered as N7614C. Her nose was reconfigured as part of flying laboratory use. By 1964, she was registered to the Pennsylvania State University and one year later to H.W. Harbican, Houston, Texas. In 1966, she was sold to Flying W. Productions, Medford, Oregon. The B-25 was modified for use as a camera ship, fitted with a hemispherical optical bubble which replaced the greenhouse nose to enable cameras to shoot through without distortion. In 1970, she was sold to John Hawke's Euramericair at Fort Lauderdale and flown to the U.K. and operated from Luton. One of her first filming contracts was in June of that year. She was to make a promotional film for BOAC, which included air-to-air assignment with their recently delivered Boeing 747. The airplane was used in Britain for a few years as a camera platform for a series of BOAC commercials. By 1973, she was at Dublin and then parked at Prestwick early in 1974. She was flown to Shoreham in 1974. Things were looking brighter, and rumours of a complete restoration abounded. But this did not happen, and she languished at the western end of the airfield. In 1976, a group of volunteers from the Duxford Aviation Society arrived at Shoreham. The B-25 was dismantled and reached Duxford at last in October that year. A replacement nose section was obtained, and the restoration could begin. In 1990, she was fitted with the B-25J glass nose and from 1997 she was displayed in the American Air Museum at the Imperial War Museum Duxford.

N7614C was used as a camera ship, with modified clear bubble nose and modified tail, used for air-to-air film work. She is photographed here at Dublin Airport in July 1973. On the nose is still written "Flight Laboratory". Note the light green painted rudders. (Collection Wim Nijenhuis)

The airplane seen at Shoreham in 1974. The nose wheel tire has already disappeared. (Richard Vandervord)

The American Air Museum, Duxford

Duxford airfield is nowadays a mecca for warbird enthusiasts and is the home base of the Imperial War Museum and the American Air Museum. The American Air Museum in Duxford is a museum dedicated to American military aviation. The museum is part of the Imperial War Museum Duxford. The IWM museum is huge, spread out over multiple hangars, and down nearly the length of the airfield. Down the end is the American Aviation Museum. Opened in 1997, this American museum, which is also a memorial to the American airmen who lost their lives during the WWII, is housed in an impressive building, designed by British architect Norman Foster to resemble a giant hangar. Inside the museum, it is possible to see the largest collection of US military aircraft outside the United States. The large collection of the American Air Museum, composed by aircraft, vehicles and artillery, including a B-17 Flying Fortress, a B-24 Liberator, a B-25 Mitchell, a B-52 Stratofortress, a SR-71 Blackbird an a Lockheed U2, among many others. The aerodrome at Duxford was built during the First World War and was one of the earliest Royal Air Force stations. Following the end of the Second World War, and once again an RAF station, Duxford entered its last operational phase and was then equipped with jet fighters. In July 1961, the last operational flight was made from RAF Duxford, and for some 15 years the future of the airfield remained in the balance. The Imperial War Museum had been looking for a suitable site for the storage, restoration, and eventual display of exhibits too large for its headquarters in London and obtained permission to use the airfield for this purpose. Cambridgeshire County Council joined with IWM and the Duxford Aviation Society, giving the near-abandoned aerodrome a new lease of life. Today, IWM Duxford is established as the European centre of aviation history. The historic site, outstanding collections of exhibits and regular world-renowned Air Shows combine to create a unique museum where history really is in the air. As mentioned, the American Air Museum was built on the site of Duxford Air Base. The American Air Museum is home to 850 objects, including equipment, uniforms, keepsakes, and photographs. Many of these have never been seen by the public. The American Air Museum was opened by Queen Elizabeth II on 1 August, 1997. The museum was re-dedicated on 27 September, 2002, in a ceremony attended by former President George H. W. Bush and by Prince Charles. On 19 March, 2016 the museum was reopened after a major re-development. The B-25 in the museum is N7614C, the former camera ship of John Hawke. In 1990, she was fitted with the B-25J nose and from 1997 she was displayed as a USMC PBJ-1J in the American Air Museum. In 2016, she received a fresh coat of paint representing a bomber of the 340th Bomb Group.

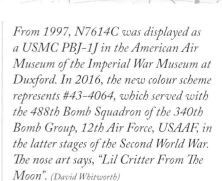

From 1997, N7614C was displayed as a USMC PBJ-1J in the American Air Museum of the Imperial War Museum at Duxford. In 2016, the new colour scheme represents #43–4064, which served with the 488th Bomb Squadron of the 340th Bomb Group, 12th Air Force, USAAF, in the latter stages of the Second World War. The nose art says, "Lil Critter From The Moon". (David Whitworth)

HANOVER STREET

As mentioned before, John Hawke was the man behind the five B-25s used in the film "Hanover Street". In this film they used five B-25s. The film is a love story and not really a war film with great combat actions. However, it has some fine action scenes. The film was directed and written by Peter Hyams and released by Columbia Pictures in 1979. "Hanover Street" was filmed in the Elstree Studios in Borehamwood, Hertfordshire. The outdoor movements and flying scenes were filmed at the airfields of Bovingdon and Little Rissington. The role of pilot David Halloran is played by a young Harrison Ford. Richard Masur played the role of 2nd Lieut. Jerry Cimino (bombardier) and Michael Sacks played the role of 2nd Lieut. Martin Hyer (co-pilot). Other actors were Christopher Plummer (famous for his role in The Sound of Music) and Lesley-Anne Down. Although B-25s were not flown from English bases by the USAAF during World War Two, the film company did not bother because there were more B-25s available at that time than any other World War Two bomber. The co-operation of the producer and the director with the assistance of Frank Tallman in the film "Catch-22" was another reason to use B-25s.

The five airplanes had the USAAF serial numbers 44-29121, 44-29366, 44-30210, 44-30925 and 44-86701. All these B-25s where J-models and TB-25s with the trainer T designation. The planes were to be picked up in Florida, Kansas, Texas and two in California. For the ferry-flight to England and the film scenes, experienced pilots were hired. John Hawke was the pilot of airplane #44-29121 with co-pilot Bill Parish. This airplane was picked up from Oakland, California. In airplane #44-29366, the pilot was David Tallichet and co-pilot Lester Eddie. This airplane was picked up from McDill AFB, Tampa, Florida. Airplane #44-30210 had on board pilot Vernon Ohmert and co-pilot Bill Baldwin. The airplane was picked up from David Tallichet's Yesterday's Air Force in Barstow, California. There she was known as the

"Tokyo Express". In airplane #44-30925, the pilot was Eric Lorentzen and co-pilot Bernhard Haddican. The airplane was picked up from San Marcos, Texas. And the pilot in the fifth airplane #44-86701 was Mike Wright and co-pilot Bill Muszala. This airplane was picked up from Topeka, Kansas where she had been sitting unattended for two years. The five airplanes crossed the Atlantic in 1978 via the Azores and they arrived on 11 May. Four together and one, #44-29366, arrived a few days later on 15 May. She had a troubled journey because of a hydraulic failure and difficulties with the ferry tank. She flew via Greenland and Iceland. The arrival of the B-25s was a cause of great excitement for the local people because it had been a very long time ago since they had seen American bombers. The preparation and the ferry-flight were documented in a tv-production called B-25 Mitchells Do Fly in IMC. This documentary was later broadcast in the U.K.

The planes were flown to Luton. Here some wooden and Perspex turrets were added to the B-25s. They emerged with realistic looking guns. The airplanes were painted Olive Drab and Neutral Grey and received their film names. Airplane 44-29121 became "Brenda's Boys", # 44-29366 became "Marvellous Miriam", #44-30210 became "Big Bad Bonnie", #44-30925 became "Gorgeous

George-Ann" and the fifth airplane 44-86701 got the film name "Amazing Andrea". Also, the airplanes were provided with very nice nose art paintings on both sides of the fuselage nose that were based on the pin-up art of Alberto Vargas. Vargas became famous in the 1940s as the creator of iconic World War II era pin-ups for Esquire magazine and known as "Vargas Girls". This is extensively described in my previous book "Mitchell Masterpieces Vol. 1". The paint on the five airplanes was water-based, so after the filming was done, it could be easily removed.

In May 1978, the airfield of Bovingdon was restored to a United States Army Air Force status for "Hanover Street". The control tower looked operational with a fresh coat of olive drab paint and the windows re-glazed. An adjacent row of Nissen huts was added to the atmosphere. A usual array of airfield vehicles was brought in, including jeeps and bomb-trains. During a few brief days at Bovingdon, the B-25s were flying on several occasions. But the emphasis was on the filming of ground sequences. By the end of May 1978, the filming at Bovingdon was complete. The airplanes went to the former

The stars of "Hanover Street"

N86427 – #44-29121

N9115Z – #44-29366

N9455Z – #44-30210

N9494Z – #44-30925

N7681C – #44-86701

RAF training field at Little Rissington were the flying sequences were filmed. During the filming "Brenda's Boys" caused the film crew the most trouble as she constantly suffered engine problems and was coughing and spluttering all the time. For the interior scenes, a B–25 fuselage was used at the Elstree Studios. They fired the machine guns, borrowed from an armoury, and fired blanks on the set. After the filming was finished, the airplanes were transferred to Blackbushe airfield in Hampshire.

Although all five airplanes were earmarked for return to the U.S.A., only one did. From Blackbushe, the airplanes went separate ways. "Brenda's Boys" was used for the film "Cuba" and went to Spain (see the chapter Spain). She is now displayed in the Air Force Museum in Madrid. "Marvellous Miriam" went to the RAF Museum in London. "Big Bad Bonnie" was bought by the Mitchell Flight of John Hawke. This was a small organisation for the rescue of Bonnie and flying air shows with the airplane. She finally went back to the U.S.A. for restoration in Chino. "Gorgeous George-Ann" was stored at several locations and is now owned by the Brussels Air Museum in Belgium where she is being restored. Finally, "Amazing Andrea" was abandoned in France and went to the Musée de 'l Air in Paris, where she was later destroyed in a hangar fire.

N86427

This B-25J-20 was delivered to the USAAF in August 1944. After use by the USAAF, she was stored at Davis Monthan AFB in 1958. From 1959, she was owned by National Metals Co, Tucson, Arizona and registered as N86427. After she had several owners, she was sold to Visionair International Inc., Miami, Florida and London, U.K. of John Hawke in 1978. On 11 May, 1978 she arrived at Luton, U.K. for use in the film "Hanover

"Brenda's Boys" at the RAF base Little Rissington, Gloucestershire, in June 1978. Like all the other B-25s from the film, she was painted Olive Drab and Neutral Grey and received a very nice nose art painting based on the pin-up art of Alberto Vargas. Her white serial number is a fake number. (Graham Salt)

Street" as #151724 "Brenda's Boys". Later she flew at a few air shows as #151451 "Miami Clipper." In January 1979, she arrived in Malaga to appear in the film "Cuba" and was painted overall yellow. On 25 January, 1979 the airplane made an emergency landing at Malaga and was abandoned. The ownership of the airplane changed in December 1984 and the airplane was restored for static display and from 1988 she was displayed at the "Museo del Aire" at Cuatros Vientos airfield in Madrid.

N9115Z

In September 1944, the B-25J-20 was delivered to the USAAF as 44-29366 and arrived at Moody Field, Georgia. She served with

Later, the B–25 flew at a few air shows as #151451 and named "Miami Clipper." (Paul Harrington)

different units until May 1958, when she was transferred to Davis Monthan AFB for storage. In January 1960, she was sold to Sonora Flying Service, Columbia, California and registered as N9115Z. She was converted into a fire tanker. She was sold again in 1964 and 1968. In 1969, she was sold to Filmways Inc., Hollywood, California, and flew in the film "Catch-22" as 6M named "Hot Pants". In 1972, she went to David Tallichet at Long Beach, California and was restored in 1977. Afterwards, she was displayed at

Three of the starring actors in front of the plane. From left to right: Richard Masur, Harrison Ford, and Michael Sacks. (Collection Wim Nijenhuis)

"Marvellous Miriam" in 1978 at Bovingdon. (Graham Salt)

From 1982, ex "Marvellous Miriam" was displayed in the Bomber Hall of the RAF museum at Hendon, London, as USAAF #34037. In 2022, she was removed from the collection and the aircraft was purchased by the Lincolnshire Aviation Heritage Centre. (Alan Wilson)

In 1980, the B-25 was renamed "Marvelous Milly" and did some air shows.
(Collection Wim Nijenhuis)

Tampa, Florida with markings of the 17th Bomb Group and flew as "Toujours Au Danger". On 15 May, 1978 she arrived at Luton, U.K. for use in "Hanover Street". She flew as #151645 and was named "Marvellous Miriam". In June 1978, she went to Blackbushe for storage. She was acquired by Doug Arnold Warbirds of Great Britain Ltd. in June 1979 and renamed "Marvelous Milly" in 1980. In 1982 she went to the RAF Museum at Hendon, U.K. At the beginning of 2022, she was removed from the collection and in August 2022, the aircraft was purchased by the Lincolnshire Aviation Heritage Centre. She is currently in museum static condition, and they hope that the future will see the B-25 roar back to life to provide aircraft experiences at East Kirkby.

N9455Z

The airplane was delivered to the USAAF in December 1944 as 44-30210 and was a B-25J-25 model. She was delivered at Randolph Field, Texas, and assigned in March 1945 to Chanute Field, Illinois. She was modified by Hayes in 1956 and stored at Davis-Monthan in 1958. In January 1960,

she was sold to National Metals, Phoenix, Arizona and registered as N9455Z. After she had several owners, she was sold to David Tallichet, Military Airplane Restoration Corp. in Chino, California in January 1975 where she flew as "Tokyo Express". In 1978, she was ferried to Luton for the film "Hanover Street", were she arrived on 11 May. In the film she flew as #151863 and was named "Big Bad Bonnie". Afterwards, she was stored at Blackbushe.

On 28 February, 1979, "Big Bad Bonnie" left England for her return flight to David Tallichet in the U.S.A. But due malfunctions in the airplane's engine systems she only got as far as Dublin. The plane landed at Dublin airport where she stayed for the next two years. After negotiations for purchase in the beginning of 1981, she was finally acquired in late April by The Mitchell Flight. This was a small organisation for the rescue of Bonnie and flying air shows with the airplane. John Hawke was the captain. The Mitchell Flight was based at Cranfield Airport, Bedfordshire. After some very exhaustive checks the airplane was airborne again for a test flight on 23 May, 1981. After the successful test flight, the airplane was flown to Liverpool Airport. John Hawke was at the controls and Rodney Small was the co-pilot. The debut of the airplane for air display should have

been 24 May at Mildenhall. But due the loss of a cylinder the flight was cancelled. One week later the problem had been solved and since that time the airplane has flown in several air shows in the U.K. and on the European continent. After that, the airplane went to Cranfield on 3 October, 1981 for a complete overhaul and new paint scheme. The owners choose to use the camouflage of the 488th Bomb Squadron of the 340th Bomb Group from Tunisia. On 18 April, 1982 the airplane made its first flight at Cranfield in the presence of James Doolittle, on the occasion of the 40th Anniversary of the Tokyo Raid. In July 1982, the plane was on her way back from a show in Switzerland, when over France the oil pressure warning light came on. An emergency landing was made at Avignon Caumont Airport. The crew was arrested, a standard procedure, and then came back by train and ferry to see if they could obtain the parts to repair "Big Bad Bonnie". Duxford did loan an engine to get her back, but there was hardly any oil pressure when it was fitted a few weeks later. The airplane then was left in France for a number of years, after John Hawke and a few others in the consortium, fell out over expenses. In 1983, the B-25 was purchased by David Tallichet and was ferried from the U.K. back to Chino, California on 1 August, 1986. In the U.S.A. she was restored and

"Big Bad Bonnie" in action during the film recordings at Bovingdon in 1978. (Richard Vandervord)

Aan engine close up taken in February 1979 at Blackbushe, shortly before her return flight to David Tallichet in the U.S.A. (Steve Fitzgerald)

After filming, "Big Bad Bonnie" was owned by The Mitchell Flight and appeared in several air shows in Europe. This photo was taken in October 1981 at Zestienhoven Airport near Rotterdam in the Netherlands. Now she has a desert camouflage scheme, and the "old" nose art was retained. (Collection Wim Nijenhuis)

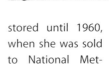

How low can you fly with a B-25? John Hawke shows it during an air show. (Collection Wim Nijenhuis)

displayed as #30210 "Big Bad Bonnie" with aircraft code 8U. She was the only airplane of the five "Hanover Street" Mitchells that finally went back to the U.S.A. Currently, she is owned by Latshaw Drilling & Exploration Co. of Tulsa, Oklahoma. A company that provides drilling services to energy exploration companies.

N9494Z

The airplane, a B-25J-30 with s/n 44-30925, was delivered to the USAAF in March 1945. She was delivered to Lubbock Field, Texas but soon moved on to the Air Training Command at Turner Field, Georgia. She was used as a trainer and later operated from different bases. In 1954, she travelled to Hayes where an extensive overhaul and conversion took place. Redesignated as TB-25N she returned to Lubbock. In April 1958, she was transferred to Davis Monthan AFB where she was

stored until 1960, when she was sold to National Metals Co, Phoenix, Arizona and registered as N9494Z. From then on she had different owners.

In December 1968, she was sold to Filmways, Inc., Hollywood, California and flew in the film "Catch-22" as "The Abombinable Snow Man" and later "Laden Maiden". In the years thereafter, she was sold again to different companies and finally in April 1978, to John Hawke's Visionair International Inc., Miami, Florida. She was ferried to the U.K. and arrived in Luton on 11 May, 1978. She was used in the film "Hanover Street" and flew as #151632 and named "Gorgeous George-Ann". Afterwards, she was stored at Blackbushe and flew later as "Thar She Blows". She was trucked from Blackbushe and later to Coventry and stored until 1994. She was then partially demolished. In 1995, she was sold to Aces High Ltd. at North Weald and registered as G-BWGR. She finished in a dismantled condition at Sandtoft. There her condition, slowly but certainly, deterio-

rated. In behalf of the Brussels Air Museum Fund, the bomber was bought and transported in 2006 to Vissenaken for restoration. In 2013, the B-25 was moved to the new association's workshop of the Belgian Aviation Preservation Association (BAPA) at Gembloux for further restoration.

After filming, she was stored at Blackbushe and later did some air shows. This picture was taken at Blackbushe in July 1978. (Bob Woolnough)

Later, the nose art of N9494Z was replaced with "Thar She Blows" and invasion stripes were added. (Richard Vandervord

N7681C

The B-25J-30 named "Amazing Andrea" was delivered to the USAAF as 44-86701 in June 1945. Initial she was assigned at Grand Island, Nebraska. In 1955, she was modified by Hayes and redesignated as TB-25N. After assignment at Wright-Patterson AFB, she was declared surplus and stored at Davis

The role of pilot David Halloran is played by the then young actor Harrison Ford. He played the pilot of "Gorgeous George-Ann", the lead plane in the film.

(Collection Wim Nijenhuis)

On 15 June, 1978 the B-25 was a guest at the air show at Deelen Air Force Base on the occasion of the 65th Anniversary of the Royal Netherlands Air Force. John Hawke is at the controls. (Wim Nijenhuis)

"Amazing Andrea" at Bovingdon in 1978. The airfield of Bovingdon was restored to a United States Army Air Force status for the film. The control tower was painted and a row of Nissen huts was added. *(Richard Vandervord)*

December 1978, the airplane at the airfield of Dinard-Pleurtuit is very weathered. The nose art and name have completely disappeared. *(Gilles Billion)*

Monthan AFB from December 1957. In April 1958, she was sold to Bud Marquis/ Airplane Crop Dusting, Marysville, California and registered as N7681C. In 1968, she was sold to Tallmantz Aviation, Orange County, and Filmways, Inc., Hollywood, California, respectively. She was used in the film "Catch-22" and flew as "Annzas". In April 1978, she was sold to David C. Tallichet Military Aircraft Restoration Group, Chino, California and ferried to Luton. She arrived there on 11 May, 1978 for use in the film "Hanover Street". In the film she flew as #151790 and named "Amazing Andrea". She was subsequently stored at Blackbushe and later in 1978 passed on to Musée de l'Air et de l'Espace in France. In December 1978, she was at the airfield of Dinard-Pleurtuix. In May 1990, she was destroyed in Musée de l'Air, Paris when the museum storage facility burned down.

ACES HIGH LTD.

Aces High was formed by Mike Woodley on 20 November, 1979 to supply aviation to the film and television industry. Mike Woodley is one of the UK's leading experts in airplanes for film and TV. His flying career initially started with the airlines, but he soon made the move to a mix of Police helicopters and vintage World War 2 fighters and bombers. He holds Airline Transport Licences for both the USA and Europe and is qualified to fly single and multi-engine aircraft, sea planes, helicopters, gyroplanes, jets and gliders. He prefers to specialise in aerial co-ordination and advising on aviation content for film and TV. In this capacity Mike Woodley has headed up Aces High since 1979. Aces High Ltd. opened its facility at North Weald in 1986 and business has

expanded ever since. The whole airfield had been rejuvenated and became a major asset to the region. Today, the company is located at Dunsfold Park Aerodrome near Cranleigh, Surrey and still run by Mike Woodley, but now under the name Aviation Filming Ltd. The company makes TV and advertising productions. It is also active in air shows and the bigger film productions. They have worked for several James Bond films, among which "Casino Royale". For this film, a Boeing 747-200, which served with British Airways until 2002 as "City of Birmingham", was purchased by Aces High and transferred to Dunsfold. It was modified and used for filming for this 2006 James Bond film. Some of the scenes set at Miami International Airport were filmed at Dunsfold. The company also played a role as aviation ground coordinator in the 1990 film "Memphis Belle" and in 2019 as aerial co-ordinator in six episodes of the TV mini-series "Catch-22". Nowadays, Aviation Filming has its own classic and modern airplanes, like a C-47 Dakota, but also Russian Mil-Mi 8 and 24 helicopters, a Huey HU-1H helicopter and private jets. If necessary, they are looking for planes and helicopters from all over the world to meet the needs of customers.

In the past, Aces High had three B-25s: the camera ship N1042B, N9089Z and N9494Z.

N1042B

One of the most remarkable B-25s in the United Kingdom was a famous camera ship previously used by Tallmantz in the U.S.A. This was the B-25J-25 with serial number 44-30823. This airplane was delivered to the USAAF in March 1945 and modified by Hayes in 1956. In 1958, she was sold to the civil market and was purchased by Wenatchee Air Services, Yakima, Washington. She was registered as N1042B. In May 1962, she was sold to Tallmantz Aviation, Santa Ana, California. She was converted into a cinema camera ship, with a camera in the nose and tail, and a camera gantry in the bomb bay that could be extended and retracted in flight. Distinctive in its modification was the construction of a new camera nose. She was used for numerous film projects including the film "Catch-22". When Tallmantz was sold to new owners in 1986, the airplane was sold to Universal Aviation, a corporation in Delaware. She was operated by Aces High Ltd. and arrived on 9 April, 1988 at the Aces High facility at North Weald

Aviation Filming Ltd. at work in 2019 for the TV mini-series of "Catch-22". *(Aviation Filming)*

in the U.K. This not only provided Aces High with another airplane for flying but, more importantly, it gave the company the possibility to use her as a camera platform. It set to work on a big contract for Aces High, filming the aerial sequences for "A Piece of Cake". Aces High cooperated very closely with its U.S.-based partner Dean Martin's Warplanes Inc. at Burlington, Vermont. The B-25 was checked out and painted in USAAF camouflage by Warplanes before she made the ferry flight to the U.K. She received an overall gloss dark Olive Drab and Neutral Grey paint scheme with yellow trim on the tail and yellow engine cowling rings as well as USAAF markings. She was named "Aces High" and later the name "Dolly" was added. Both were painted on the nose. Aces High then owned two B-25s, the new camera ship and the old "Bedsheet Bomber" that was purchased in 1983 following the sad demise of the Historic Airplane Museum at Southend. Aces High used the B-25 as a camera ship, and she was used specifically to film "Memphis Belle" in 1989. Tony Ritzman of Aero Trader was one of the camera pilots hired to fly the airplane. After completion of the filming, the airplane was seen on several air shows in Europe and later her camera nose was removed and replaced with the traditional greenhouse nose. In 1996, the airplane went back to the U.S.A. after she had been sold to World Jet Inc. at Fort Lauderdale, Florida. There she was overhauled by Tom Reilly and continued to fly until today. Nowadays she flies with the name "God and Country".

The former ship of Tallmantz at Duxford in May 1988. The airplane was overall gloss dark Olive Drab and Neutral Grey with yellow trim on the tail and yellow engine cowling rings as well as USAAF markings. She was named "Aces High". Her registration N1042B is painted black on the vertical tail. (Andy Robinson)

Southend Historic Aircraft Museum

The Southend Historic Aircraft Museum was located on Aviation Way, opposite the County Hotel. Before the Southend Historic Aircraft Museum (SHAM) opened in 1972, a small group of aviation enthusiasts who were based at Biggin Hill airport in Kent, formed a small collection under the name of The British Historic Aircraft Museum.

Their main task was the restoration of the B-25 HD368 that had been damaged during the Biggin Hill Air Fair in 1966. The group moved to Southend in 1967 and made an application to construct a hangar to display the airplanes they had collected. Local aviation enthusiasts were recruited to work in their spare time on the renovation of the airplanes. The project collapsed shortly after the application had been approved and the airplanes, including the B-25, were left on the eastern boundary of the airport. When it seemed that the airplanes were destined for the scrapheap, a consortium of businessmen stepped in to save them. The consortium members who were all aviation enthusiasts and amateur pilots, formed the company that took over the ownership of the airplanes and breathed life back into the museum project. Building work started with the preparing of the site for a museum and once the main exhibition hall had been completed, the airplanes started to move at the site during 1970. In February 1972, with work nearing completion on the museum site, the first two airplanes were moved from the storage facility on the airport to the museum complex. The museum was opened on 26 May, 1972 by Air Marshal Sir Harry Burton KCB, CBE, DSO, RAF who at the time was Air Officer Commanding-in-chief of Air Support Command. The following day the museum was opened to the public and an air show was held to celebrate the completion of the project. The museum, which had started off as a very popular attraction, slowly started to fall from popularity and in 1983, after 10 years of hard work the Historic Aircraft Museum at Southend closed on 27 March. The closure of the museum was put down due to a number of causes, including rising costs, falling number of visitors, the opening of the much larger RAF Museum at Hendon and poor support from the council. On 10 May, 1983 an auction was held at the museum to sell off the airplanes and the rest of the museum's collection of aviation artefacts.

N9089Z

This B-25J-25 with s/n 44-30861, was delivered to the USAAF in March 1945 and assigned to numerous units. In May 1954, she was modified by Hayes and was assigned to Lackland AFB, Texas. After four years she was stored at Davis Monthan, Arizona. In December 1959, the airplane was acquired by the American Compressed Steel Corporation from Cincinnati, Ohio, and would be operated by the Aero American Corporation under the control of Crewdson's Film Aviation Services in the U.K. on behalf of Columbia Pictures. She was registered as N9089Z. The airplane arrived in the U.K. in 1963 and was put to work filming "The War Lover" and "633 Squadron". The airplane was named "Moviemaker II" and was modified as a camera ship with a flat nose panel. In the film "633 Squadron" she played a part on both sides of the camera. Not only as the camera ship, but also in a role in front of the cameras at Bovingdon and painted up in pseudo-RAF markings as N908. After its use in "633 Squadron" the airplane was laid up at Biggin Hill because of a large and unpaid customs bill.

In 1966, she was sold to the British Historic Aircraft Museum. From Biggin Hill, she was moved to Southend in 1967 by a crew of aviation enthusiasts. The spar bolts would not loosen, so it was impossible to remove the wing centre section. Apparently, the crew foreman decided that the best way to solve the problem was to take a chainsaw to cut the main spar, ensuring that she would never fly again. After the museum became the Southend Historic Aircraft Museum and moved to the compound on Aviation Way, hundreds of hours were spent on the B-25 and the production of very realistic wooden guns for all the gun positions. The B-25 used to be in a fair static condition when it was at the Southend Historic Aircraft Museum. With a somewhat unusual RAF colour scheme and a full set of guns, she was displayed as HD368 with the aircraft code VO-A. During the war, the British Mitchells were normally Olive Drab/Neutral Grey or Dark Green/Dark Sea Grey. Only a few very early Mitchells had the three-tone camouflage like the Southend Mitchell. The flat front camera nose was replaced with a bomber example quite early on.

When the other B-25 camera ship N7614C of John Hawke was in the U.K. to film a BOAC Boeing 747 promotional film, she suffered from a problem with the nose wheel leg and the owners managed to borrow N9089Z's nose wheel leg for a while to complete the filming, leaving N9089Z propped up on oil drums for a while. After the BOAC filming was completed, N7614C was abandoned at Shoreham and was eventually passed on to Duxford for preservation. N9089Z was repainted a second time as HD368 before the Southend museum opened in 1972. During the repaint a fair bit of the underwing skin was replaced after removal of masses of birds' nests and much of the airframe was stripped to bare metal.

Things started to go downhill when the top turret was "borrowed" for the London

1963, N9089Z seen at Luton Airport. Still with her greenhouse nose. Later she was registered as G-BKXW and operated for a while by The Fighter Collection at Duxford. (John W. Read)

Weekend Television (LWT) production "We'll meet again" for use in producing a top turret for the B-17 "Sally B". She was never returned, hence the crudely plated over turret hole. After the auction in 1983, she stood for months at the closed museum before moving to Duxford and into the "care" of Aces High. She was moved to the museum site during 1984 and later entered the British Aviation Register as G-BKXW. She then got damaged in high winds before departing to the new base of operations at North Weald. A repair paint job was applied, and she became "Bedsheet Bomber". She was presented in the new paint scheme in April 1988. In 1990 she was sold to the Fighter Collection at Duxford and was dismantled and stored. In February 2006, she was stored at Wycombe Air Park, also known as Booker Airfield, awaiting restoration.

In July 2019, the cockpit section was delivered to the Wings Museum near Balcombe,

West Sussex. Their work is focussing on the cockpit section of "Bedsheet Bomber". The rest of the airframe is in storage. The Wings Museum also has a second cockpit section of another B-25J which was salvaged from the Aleutian Islands and is under long term restoration.

The Wings Museum was established in 2003 by the brothers Hunt. Sharing a lifelong passion for warbirds for over 40 years and with a growing collection of memorabilia, the Hunt brothers established the museum. The museum was based at the "Old Gas Decontamination Block" at Redhill Aerodrome. Later the museum moved into Hangar 9 at Redhill Aerodrome. The museum has since relocated to Balcombe in West Sussex. The Wings Museum is housed in a large hangar style building and the displays are broken down into dedicated subject areas. At the centre of the museum is a complete fuselage from a Douglas C-47 Skytrain which visitors can walk inside. This was used during the filming of the TV hit series "Band of Brothers".

In 1966, she was sold to the British Historic Aircraft Museum that later became the Southend Historic Aircraft Museum. At Southend, she was painted as No. 98 Squadron Mitchell HD368 with the squadron code VO-A. Here she is at the Southend Historic Aircraft Museum in 1972. She has an unusual RAF colour scheme. The flat front camera nose was still present at the time. (Richard Vandervord)

After filming "633 Squadron" and other productions, she was withdrawn from use circa 1964 and seen here at Biggin Hill. On her tail is painted the name "Aero Associates". (Collection Wim Nijenhuis)

Left: *The ship photographed in June 1978. The flat nose has been replaced with the usual B-25 greenhouse nose.*
(Paul Harrington)

After the museum auction in 1983, she was moved to Duxford and later North Weald. A repair paint job was applied, and she became "Bedsheet Bomber". She was presented in the new paint scheme in April 1988. Here she is at North Weald in 1990.
(Malcolm Clarke, Robin Edridge)

Centre: *Unfortunately, she was dismantled and stored in February 2006 at Wycombe Air Park, also known as Booker Airfield, awaiting restoration.* (Alan Allen)

Bottom: *In July 2019, the cockpit section was delivered to the Wings Museum near Balcombe, West Sussex. In early October, they took advantage of some good weather and got the entire cockpit exterior blasted back to bare metal. This meant removing the many layers of paint and giving the possibility of getting a clean, corrosion-free surface ready for the yellow primer. At right, the result of the cockpit section of N9089Z at the end of 2019. The overall paint scheme planned for the near future is that of an RAF No. 98 Squadron Mitchell.*
(Wings Museum)

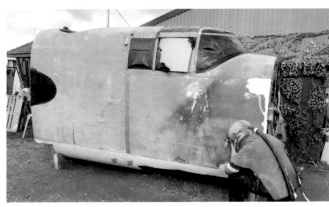

N9494Z

The B-25 "Gorgeous George-Ann" from the film "Hanover Street" was stored at Blackbushe after filming. She was trucked from Blackbushe and later stored at Coventry. She was then partially demolished. In 1995, she was sold to Aces High Ltd. at North Weald and registered as G-BWGR. Aces High, however, owned the airplane only a few years. She was stored at Coventry Airport where she was parked at a place called "Rock Farm". In 1998, she went to David Copley/Imperial Aviation Group at North Coates. She ended up in a dismantled condition at Sandtoft. Here she stayed until 2006, when she was bought by the Brussels Air Museum Fund (BAMF) and was transported later that year to Belgium.

A sad sight. This picture of N9494Z was taken at Coventry in February 1995. The airplane got further demolished. (Chris England)

The two former pin-ups of "Laden Maiden" and "Gorgeous George-Ann" are still visible. Even the traces of former owner John Stokes from San Marcos are still there. (Chris England)

THE FIGHTER COLLECTION

The oldest civil B-25 in Great Britain was a B-25D-30 with s/n 43-3318. This airplane was displayed and flown for several years with The Fighter Collection at Duxford. After her military service with the RCAF, she was registered as N88972 in the U.S. and had several owners. In 1967, she went in active service as a civil fire bomber. In May 1980, she was bought by Merrill Wein from Anchorage, Alaska. He spent seven years working on the airplane until she was moved to Aero Trader in Chino for restoration. In 1987, the airplane was sold to The Fighter Collection, which worked with Aero Trader to complete the restoration and took her to England for a career as a warbird classic. After restoration, she flew from St. John's in New Foundland to Luton on 7 November, 1987 to arrive at Duxford the very next day. She was painted with the RAF serial KL161 and aircraft code VO-B of No. 98 Squadron and named "Grumpy". In March 1999, she

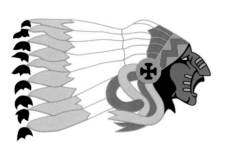

FLYING LEGENDS

Pictured in January 1988, N88972 of The Fighter Collection. The ex-RCAF airplane arrived at Duxford in November 1987. She is overall natural aluminium finished and still has to receive her nice British WWII camouflage pattern with the "Grumpy" nose art. (Andy Robinson)

was entered in the U.K. registry as G-BYDR. She was displayed in the museum's hall at Duxford and did several air shows in the U.K. and on the European continent. The B-25 moved on from The Fighter Collection in 2003, when she was placed in storage at North Weald. In 2008, she was restored to an airworthy condition and returned to the U.S.A. in 2009. Nowadays she is a famous warbird of the Historic Flight Foundation at Paine Field, Washington.

The Fighter Collection (TFC) is a private operator of airworthy vintage military aircraft or warbirds. It is based in the United Kingdom at Duxford Aerodrome in Cambridgeshire, an airfield that is owned by the Imperial War Museum and is also the site of the Imperial War Museum Duxford. The Fighter Collection was founded by Stephen Grey in 1985. They own the largest collection of airworthy WWII warbirds in Europe. TFC is run by a professional team of engineers and pilots, who do all the work at air shows and for film work. TFC included many unique examples of aircraft designed and produced during the pioneering days of aviation through the 1930s and 1940s with aircraft taking part in air shows and starring in film and movie work through Europe and the United Kingdom. TFC also runs the greatest air show in Europe each year, the

"Grumpy" photographed in the hangar at Duxford in October 1989. During the war, one of the Mitchells of No. 98 Squadron was named "Grumpy" after one of the seven dwarfs in Disney's 1937 film "Snow White and the Seven Dwarfs".
(Wim Nijenhuis)

This is "Grumpy" on her way at an air show at Eindhoven Air Force Base in the Netherlands in April 1995. She is provided with Invasion stripes and on her right fuselage side she has the Indian head of The Fighter Collection. *(Wim Nijenhuis)*

Flying Legends Air Show at Duxford airfield. A world-famous air show of great classic piston engine fighters, bombers and legendary aircraft of both First and Second World Wars.

Right:
Duxford, July 2009. The bomber now has the large number N88972 on her fuselage and the Indian painting on her nose has been removed. Later that year, she returned to the U.S.A. (Cor van Gent)

Details of the nose art on both sides of the fuselage.
(Collection Wim Nijenhuis)

Netherlands

In the Netherlands there are still three B-25s. Two are military airplanes and are displayed as static items in the National Military Museum at Soesterberg and the War Museum at Overloon respectively. The third one is a civil B-25J-20, s/n 44-29507, with the former U.S. registration N3698G. In June 1981, she was sold to Donald Webber of Aerial Solutions Inc. at Baton Rouge, Louisiana. She flew as "Cochise". In June 1989, the airplane was acquired by Edwin Boshoff, Maarten van Eeghen and Joe Hartung, founders of the Duke of Brabant Air Force (DBAF), Netherlands. The B-25 was moved to Amho Corp., Wilmington, Delaware and in 1990, the registration was changed to N320SQ (representing No. 320 (Dutch) Squadron) of the RAF. During World War Two, this squadron flew with Mitchells. In May 1990, she was ferried via Canada, Greenland, Iceland and Scotland to Eindhoven Airport, the Netherlands where she arrived on the 25th. After arrival, she went to a maintenance hangar at Schiphol, were she got a new camouflage scheme. This was a WWII RAF scheme of No. 320 Sqn. with the aircraft code NO-V and serial HD346. She was named "Lotys II". Since the airplane is a J- model, this scheme was chosen because it was carried in the war by a B-25J model. However, this airplane had no top turret and the fuselage gun packs below the cockpit where fake and of the wrong shape. The bomber was operated

A very weathered "Cochise" in 1990, before departure to Europe. (Collection Wim Nijenhuis)

After arrival at Eindhoven on 25 May, 1990, she was flown to Schiphol for a new paint scheme. Here she is standing in a hangar at Schiphol with the new green/grey camouflage. The windows are still taped, and the markings still must be applied. *(Wim Nijenhuis)*

by the Duke of Brabant Air Force (DBAF) at Eindhoven, which owned the bomber. From 1990 to 1999, she flew in the RAF colours. In 1999, she was repainted in former Royal Netherlands East Indies Army Air Force colours and renamed "Sarinah" with number N5-149. In 2010, her civil registration was changed to the current PH-XXV. Ownership was transferred in 2010 to the Royal Netherlands Air Force Historical Flight after a merger with the DBAF.

Since 1990, this bomber flies regularly at European air shows and events. The Royal Netherlands Air Force Historical Flight operates an impressive fleet of historical aircraft and all in airworthy condition. The fleet mainly comprises aircraft that have flown in the past with the Royal Netherlands Air Force and Navy. The Historical Flight is a voluntary organisation consisting of aircraft mechanics, experienced pilots, and many volunteers. They operate from their home base Gilze-Rijen Air Force Base. Its origins date from 1969 when historic aviation started at Gilze-Rijen and eventually led to the current foundation. During the past few decades, the foundation has grown into a leading aircraft museum with a unique collection of airworthy historical military propeller-driven aircraft.

The B-25 was painted in a WWII RAF scheme of No. 320 Sqn. with the aircraft code NO-V and serial HD346. She was named "Lotys II". *(Wim Nijenhuis)*

After a long period of hard maintenance work, it was time for an engine run. A fine action shot taken at Gilze-Rijen in August 2019. (KLuHV)

BELGIUM

Nowadays, there is one B-25 under restoration in Belgium. This is N9494Z, the B-25 "Gorgeous George-Ann" from the film "Hanover Street", see chapter United Kingdom. After filming, she changed hands several times and was stored at various locations like Coventry Airport. She finished in a dismantled condition at Sandtoft. There her condition, slowly but certainly, deteriorated. In December 2004, one of the members of the BAMF discovered in "Flypast" news that the B-25 N9494Z stored near Hull was on sales after demise of the Imperial Aviation Group.

In 1999, she was repainted in former Royal Netherlands East Indies Army Air Force colours and renamed "Sarinah" with a bold nose art and number N5-149. On the right side, she had the name "De 2Cs". Her civil registration was changed to PH-XXV. The ship is photographed here at the June 2008 air show at Leeuwarden Air Force Base. (Wim Nijenhuis)

"Sarinah" with the Dutch flag as national markings at her home base Gilze-Rijen in May 2014. (Wim Nijenhuis)

In 2006, the fuselage and smaller parts of N9494Z were transported to a temporary storage location at Vissenaken in Belgium. The aim of the Belgian Aviation Preservation Association (BAPA) is to restore the bomber as "Pat's Victory", a RAF Mitchell of No. 320 (Dutch) Squadron and coded NO-V. (BAPA, Collection Eric Dessouroux)

Members of the BAMF discover the remnants of N9494Z at Sandtoft. (BAPA)

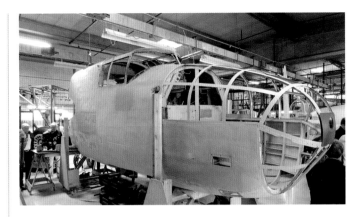

In 2013, the B-25 was moved to the workshop of the BAPA at Gembloux for restoration. This is the partially restored nose in September 2015. During the Open Hangar Day of the BAPA on 31 August, 2019, the nose section was provided with a temporary nose art "Yankee Doodle" from the TV mini-series "Catch-22". *(Wim Nijenhuis, BAPA)*

In 2005, a small Belgian project team was created and studied the possibility to raise funds for the project to buy and transport N9494Z to Belgium. Just before the New Year 2005-2006, the Brussels Air Museum Fund (BAMF) gave her formal approval to house the administrative and financial aspects of the project. A bid was placed, which was accepted. The airplane arrived in Belgium on 23 May, 2006. The airplane was stored at Vissenaken for restoration. From 2008, the paint was removed on the airframe and anti-corrosion processes were applied sparing years of manual operations and permitting a safe storage of the aircraft. In 2010, an agreement was found with the airfield at Grimbergen to store the aircraft. Unfortunately for environment reasons, it was impossible to start restoration activities there. A quest for a suitable location and spare parts then started. In 2011, one of the volunteers of the original B-25 team acquired a workshop in an industrial building in Gembloux in the heart of one of the most dynamic regions of Belgium. A big task started to setup the place ready to welcome the Mitchell. To facilitate future restoration activities and avoid misunderstanding, the volunteers and the BAFM decided in 2012 to setup a new organisation focusing on restoration and heritage preservation. The BAMF continue her activities in fund raising but the ownership of the B-25 was transferred to the Belgian Aviation Preservation Association (BAPA) on 31 December, 2012. The BAPA association targets are the restoration, conservation, exhibition of aircraft and any activities relating to the preservation of aviation heritage in any form in Bel-

gium. Their main project is the restoration of the B-25. In 2013, the B-25 was moved to the new association's workshop of the BAPA at Gembloux for further restoration. The BAPA is still looking for parts, tools, money, and motivated volunteers to work on this project. The aim of the BAPA is to restore the bomber to static display as "Pat's Victory" in reference to a RAF Mitchell of No. 320 (Dutch) Squadron and coded NO-V.

FRANCE

Around the nineties of the last century, France had one civil B-25. This was a B-25J-35 with s/n 45-8811 and in the U.S. registered as N9621C. In May 1991, the Mitchell arrived in Europe. She was sold to Apache Aviation at Dijon, France, and operated by Flying Legends, Dijon. In September 1992,

The Mitchell of Lafayette Aviation was registered as F-AZID and painted in an attractive desert camouflage scheme as was used by the USAAF during World War Two in the Mediterranean Theatre. She is seen here at an air show at Le Bourget, Paris, in May 1994 with the letters HD on her vertical tails. *(Wim Nijenhuis)*

In 2004, she was transferred to Switzerland, but returned to France again in 2008. Back under French wings, she got the registration F-AZZU and the letters SB with the number 458811 on her tail surfaces. The nose art applied by the Swiss was retained.
(Dré Peijmen, Cor van Gent)

Some details of the nose art and the tail. (Collection Liliane Cotton, Collection Wim Nijenhuis)

she was bought by Franklin Devaux of Lafayette Aviation, was registered as F-AZID and stayed at Dijon-Longvic. As F-AZID she was operated by Lafayette Aviation for several years in the European air show circuit. She was completely restored, flight controls renewed, wings removed and checked and treated all over with anti-corrosion products. She was painted in an attractive desert camouflage scheme as was used by the US-AAF during World War Two in the Mediterranean Theatre. In 2004, she was acquired by Aiject Ltd. in Switzerland where she would join Jet Alpine Fighter at Sion. But in 2008, she returned to France again. She was acquired by the French Société de Développement et de Promotion de l'Aviation based at La Ferte Alais, France and then she was registered as F-AZZU. She continued to fly as "Russell's Raider", a paint scheme that was applied by the Swiss. On 31 May, 2011, she was badly damaged by engine fire after a forced landing at Melun-Villaroche, France. The crew was performing a local flight around Melun-Villaroche Airport. A few minutes after takeoff, it seems that the right engine fired. The pilot tried to make an emergency landing, but the Mitchell hit power cables and crashed in flames in an open field. Both occupants were uninjured, but the aircraft was seriously damaged on the fuselage, wings, and engines.

SWITZERLAND

As mentioned before in the French chapter, the B-25 registered as F-AZID was acquired by Aiject Ltd. in 2004. In Switzerland she would join Jet Alpine Fighter at Sion and was registered as HB-RDE. After a period of inactivity, she was completely overhauled and finally delivered to Jet Alpine Fighter which operated her from their home base Sion in Valais. During the following years, she was operated from Switzerland until 2008 when she was acquired by the French Société de Développement et de Promotion de l'Aviation based at La Ferte Alais and returned to France. Jet Alpine Fighter was founded in 2002 under the impetus of the Valaisan Stéphane Brugnolo. He started flying at the Sion Engine Flying Group. He has been present for many years in the international air shows circuit. His experience of flying in the mountains and on vintage aircraft allowed him to become aerial coordinator of several movies or soap operas. The idea was to fly in the heart of the Alps military jets, previously used in the James Bond movie, hence the name "Jet Alpine Fighter". With a T-28 Trojan the JAF took its quarters in Sion, and the B-25 was based here and was a success in the European circuit.

AUSTRIA

Since 1995, a B-25 joins the existing fleet of historic aircraft of the "The Flying Bulls" in Austria. This airplane is a B-25J-35, s/n 44-86893, and was registered in the U.S. as N6123C. In the U.S. she flew until 1993 as "Fairfax Ghost". In 1994, Sigi Angerer, former Chief Pilot of The Flying Bulls, found her in Texas and he bought her the same year. She was completely refurbished in the U.S. and converted into a civil version with comfortable interior design in about 20,000 hours of work. She was stripped of all her paint and polished bare metal would become

Details of the tail of the Mitchell with the Swiss registration HB-RDE in small letters below the horizontal stabiliser. (Dariusz Kunicki)

The Russell's Raiders group in their Swiss environment. (Jet Alpine Fighter)

her trademark in the future. On 31 July, 1997 she was acquired by Red Bull Aviation Inc., Las Vegas, Nevada. In September 1997, the aircraft was ready for take-off to cross the North Atlantic and she was flown from Breckenridge, Texas to Salzburg in Austria. The year 1999 led to the founding of the company The Flying Bulls. The technically and visually perfect aircraft of the Flying Bulls have since then been welcome air show participants and an attraction in every type of aviation event. By the late 1990s, there was no longer enough space for the rapidly expanding Flying Bulls fleet, originally based at Innsbruck Airport. This led to a plan to erect a hangar at Salzburg Airport. The time had also come when the heretofore loose association of pilots and mechanics needed to be brought under a common roof. At the beginning of 2001, the B-25 was moved to Salzburg Airport. In 2003, the airplanes of the Flying Bulls moved into what is certainly a spectacular place, Hangar-7 at Salzburg Airport. This is a unique architectural work of art. Originally planned to accommodate the Flying Bull's collection of historic aircraft, Hangar-7 has become synonymous with avant-garde architecture, modern art and exquisite dining. Now the Mitchell makes a fine figure, whether serving as a photographer's model in Hangar-7 or taking part in one of the many great air shows in which she flies.

Polished bare metal and the Red Bull logo are a trademark of the Flying Bulls' B-25. (Cor van Gent)

In 2018, the nose art was provided with a pin-up. (Šimon Birský)

Detail of the nose art and of the vertical tails with U.S. flag and the registration number N6123C. Note the specially shaped glass cone at the tail. (Collection Wim Nijenhuis)

The polished bare metal B-25 is a true master-piece. In 2003, model manufacturer Revell even produced a 1/48 plastic scale model as a gift set. The set consists of the airplane kit, including basic paint set, brush, glue, and thinners. There is also a can of Red Bull included, plus a B-25 poster and a multi-media CD-ROM, which takes one on a virtual tour of the original aircraft's interior. In addition, there are breath-taking photos of the aircraft in flight. The spectacular pictures are supplemented by original aircraft sounds. (Flying Bulls, Collection Wim Nijenhuis)

On 25 January, 1979, the airplane made an emergency landing at Malaga after she hit an obstruction in very low level flights during the filming. So the B-25 was abandoned at Malaga. After long negotiations, the ownership of the airplane changed in December 1984 from the U.S. owners to the Spanish Air Force Museum (Museo de Aeronáutica y Astronáutica o Museo del Aire). She was dismantled at Malaga in January 1985 and trucked to the museum in Madrid. She was restored for static display and from 1988 she was displayed at the museum and painted in desert camouflage colours like the military B-25D-10 #41-30338 with call sign 74-17. The "Museo del Aire" museum at Cuatros Vientos airfield in Madrid houses an impres-

In January 1979, the yellow painted airplane N86427 was abandoned at Malaga, Spain after having been damaged during filming of the film "Cuba". The airplane was painted overall yellow. "Cuba" was a film from the MGM Studios Inc. directed by Richard Lester in 1979. The film starring Sean Connery, Brook Adams and Denholm Elliot reconstructed the life and times of Cuban dictator Fulgencio Batista.
(Collection Wim Nijenhuis, Francisco Andreu)

SPAIN

Spain was not a country that used the B-25. However, during World War II, Spain interned several airplanes, both Allied and Axis and one interned B-25 served in Spain for a few years. This was a VIP transport aircraft operating for the RAF and was Spain's only military B-25. She was a B-25D-10 with s/n 41-30338 and had the Spanish registration number 74-17. In 1953, due to the lack of spare parts, the B-25 was retired from service and finally auctioned for scrap in

1956. Spain also had one civil B-25. In 1979, this airplane made an emergency landing at Malaga and the plane can still be seen to this day in the Air Force Museum in Madrid. This is a B-25J-20, s/n 44-29121, and was registered as N86427. From April 1978 to 1984, the ship was owned by John Hawke of Visionair International Inc., Miami, Florida and London, U.K. On 11 May, 1978 she arrived at Luton, U.K. for use in the film "Hanover Street" as #151724 nicknamed "Brenda's Boys". Later she flew at a few air shows as #151451 "Miami Clipper". On 10 January, 1979, she arrived at Malaga to appear in the film "Cuba" and was painted overall yellow.

sive collection of more than 100 aircraft. In December 1948, the first studies were done to see whether a museum could be created in de new building of the Ministry of Aviation. Finally, after years of considering the Museo de Aeronáutica y Astronáutica was created on 16 June, 1966. When the museum got several donations including a series of historic airplanes in 1969, they had a huge problem as they did not have room for these aircraft. They needed a bigger area and the aircraft needed their own building. However, they still did not have a location where to establish the museum. Finally, in 1975, the airfield of Cuatro Vientos, the air-

The weathered N86427 of the Spanish Air Force Museum in Madrid around 2015. In the late 1980s, she was restored for static display and painted in camouflage colours like the interned military B-25D-10 in 1944, with call sign 74-17. (Collection Wim Nijenhuis, Contando Estrelas)

field where Spanish Aviation started, came out as the best option. On 24 May, 1981 the Museo del Aire opened its doors to the public for the first time. The objective of the museum is to acquire, conserve and display the aircraft, equipment and associated paraphernalia that constitute the historical heritage of the Spanish Air Force. It has an exterior exhibition and seven hangars. They house aircraft, uniforms, medals and decorations, vehicles, mock-ups, and other aviation-related displays. Following the end of the Spanish Civil War and the creation of the Air Force, the idea was conceived to create a museum that would reflect the evolution and history of the Spanish aeronautics.

AUSTRALIA

Australia had three B-25s on her continent. One is a military airplane and is an old D-10 model with s/n 41-30222. This ship is named

"Hawg-Mouth" and is partially restored and displayed at the Darwin's Aviation Museum/Australian Aviation Heritage Centre at Darwin. The other two are B-25s with a civil registration. One is owned by Reevers Warbirds at Adelaide and the other was owned by the Australian War Memorial.

REEVERS WARBIRDS, ADELAIDE

Reevers Pastoral Pty Ltd. is a privately owned company. Initially set up to research reverse breeding of beef cattle, but it has diversified into the warbirds market, essentially utilizing the same research techniques, and demonstrating the effectiveness of its capabilities. While travelling for agistment and breeding, the owner's interest in aircraft was piqued when stumbling across wrecks, parts, and stories from farmers. Today, while Reevers still maintains an active interest in cattle, airplanes now form a major component of the company's efforts. Owner Peter R. Smythe has been a warbird enthusiast, researcher, and collector for a long time, with

The nose section of "Lucky Lady" on a truck in Australia. She still has the nice lady nose art. (Reevers Warbirds Roundup/News)

his efforts resulting in the ever-expanding Reevers Warbirds collection. This collection has grown into a diverse and unique fleet for Australia. One of its airplanes is the B-25 with s/n 44-31508.

This B-25J-30 was delivered to the USAAF in June 1945. After her USAF service, she was registered as N6578D and sold to several U.S. owners and finally went to the United Kingdom. There she was used as camera ship for the film "Battle of Britain". This is described in the chapter United Kingdom. In January 1969, after completion of the filming, the airplane returned to the U.S.A. She was abandoned but after restoration by Tom Reilly, Kissimmee, Florida, she became airworthy again in 1981. She was flown as "Chapter XI" and later as "Lucky Lady". In

1998, she was withdrawn from use and placed in open storage at Franklin, Virginia and degraded to derelict condition and missing an engine. In April 2015, she was purchased by Reevers Warbirds in Australia. From late April to early May 2015, the disassembly of the B-25 took place and she was being prepared for shipment to her new home in Australia. The tear-down team did a great job, but it became apparent during the work that the airplane would need some major repairs once in Australia to slow down some serious corrosion problems the airplane suffered from. Once in Australia, a complete inspection took place to identify the missing parts. The long-term plan of Reevers was to restore the B-25J to flying status.

Reevers unveiled their B-25 restoration project on 8 April, 2017 at Parafield Airport, Adelaide, South Australia. The B-25 has been painted as "Pulk" (which was originally N5-131, a B-25C from No. 18 Squadron Netherlands East Indies Army Air Force). The original "Pulk" was named after the B-25's Dutch pilot, Fred "Pulk" Pelder. The accurate nose art on Reevers' B-25 was made by the son of the pilot also named Fred Pelder.

Australian War Memorial, Canberra

The bomber was registered as VH-XXV and painted as RAAF No.2 Squadron A47-31 with the code KO-P. The ESSO tiger on both sides of the nose was maintained. Here she is in February 1985 on display at the Tullamarine Open Day, Melbourne (top), and at the Mangalore Air Show in April 1985 (left).

(John Richard Thomson, Gavin Hughes)

The B-25J-30 with s/n 44-86791 was delivered in July 1945. She was put into storage in the late 1950s, until used in the 1960s as an aerial tanker for firefighting in Alaska. In the 1970s, she headed for California for movie and TV work. She was registered as N8196H. In August 1979, she was sold to Donald Gilbertson of Fairbanks, Alaska. In 1983, she was sold to Aero Heritage Inc. of Melbourne, Australia. She was restored by Aero Nostalgia of Stockton, California and ferried from the U.S. to Australia in December 1983. Aero Heritage came to an arrangement with the Australian War Memorial for the AWM to take over ownership of the B-25, which would be flown by Aero Heritage in RAAF markings for three years, prior to eventual museum display at Canberra. She was registered as VH-XXV and flew as RAAF Mitchell A47-31 with the code letters KO-P until 1987. For four years, she made several air show appearances. Unfortunately, this superb example of Australian WWII heritage fell victim to AWM politics. When her flying period ended in November 1987, she was stored because AWM management deemed her "unrepresentative for Australia" and sold her back to the U.S.A. After her last flight on 30 November, 1987, she was put in long term storage at the Treloar Technology Centre at Canberra. The aircraft was flown to Canberra, the outer wings removed, and stored out of public sight in a storage building. In March 1999, she was purchased by Yanks Air Museum at Chino, California. She was shipped from Sydney to her new home in Chino to undergo an extensive restoration. In May 2002, her current registration was issued as N6116X. Her first flight after restoration was in June 2002.

Soviet Union

From 1941 until 1945, the United States and the Soviet Union ferried about 14,000 warplanes from the U.S. to the Soviet Union. A total of 865 of these airplanes were B-25s and assigned to the Soviet Air Forces (Voyenno-Vozdushnye Sily - VVS). These B-25s were B, C, D and J models. The Soviet Union received also at least six B-

25Gs. During the war, the Soviet Mitchells were predominantly used in the Air Armies of the VVS. But because of the good flying characteristics and strong structure permitting extensive modifications, the B-25 was a very popular testbed for various research

CCCP-A370 was a B-25J-5 with U.S. s/n 43-27860 and used by Aeroflot. The colours seem to be still Olive Drab over Neutral Grey. On the fuselage she has the registration CCCP-A370. On the vertical stabiliser she has an unidentified number in small figures. (Collection Wim Nijenhuis, Aviaforum.ru)

and development projects in the late 1940s. In the Soviet Union, B-25s were used in tests from 1947 until 1949, like flight testing of early jet engines, development of air-to-air tanking methods, testing of rocket-assisted take-offs, catapult experiments and several B-25s were also converted to VIP-transports.

With the end of the war and the expiration of the Lend-Lease programme, the Soviet Union was obliged to return the airplanes to the Americans. Usually this was reduced to destruction of technology under control of American representatives. In 1945-1947, part of the machines was indeed destroyed under the supervision of inspectors from the U.S.A. However, a certain number of

Two pictures of unarmed B-25Js with civil designation used by Aeroflot. The name Aeroflot is painted on the nose. It is unknown whether the planes on both pictures are the same. Aeroflot is one of the oldest airlines in the world, tracing its history back to 1923. During the Soviet era, Aeroflot was the Soviet national airline and the largest airline in the world.
(Collection Wim Nijenhuis)

Dobrolet / Aeroflot

In the 1920s, after the end of World War I, European countries increasingly used aviation for peaceful purposes; transporting passengers, mail, and cargo. The Soviet Union kept pace with its neighbours with flights abroad, generally operated using re-equipped warplanes. On 1 May, 1922, flights on the first international route Moscow-Konigsberg (part of Germany at that time), were launched and later extended to include Berlin. 9 February, 1923 is considered the official birth date of Russian civil aviation, when the Labour and Defence Council issued a resolution to assign the technical supervision over the airlines to the Chief Administration of the Air fleet and organise the Board of Civil Aviation. Following this resolution on 17 March, 1923, the Russian Society for Voluntary Air Fleet Dobrolet was established. It was the nation's first major civil air organisation. The main goal of this commercial organisation was to develop civil aviation in the country for the needs of national economy. 15 July, 1923, marked the introduction of the first regular domestic route between Moscow and Nizhny Novgorod. Dobrolet expanded through the decade, as it extended its service to far off places into Siberia and even Outer Mongolia. But the Communist leaders were out to eliminate all private property. As a result, on 25 February, 1932, all civil aviation activities were consolidated under the name of Grazhdansky Vozdushny Flot (GVF), Civil Air Fleet, and on 25 March, 1932 the name "Aeroflot" was officially adopted for the entire Soviet Civil Air Fleet. The company is one of the oldest airlines in the world. In 1991, after the collapse of the Soviet Union, former Soviet republics and regions in Russia began founding their own airline companies. Today's company is the successor to the Aeroflot name and trademark of the former Soviet air carrier. In June 1991, the "Aeroflot Soviet Airlines" Commercial Production Alliance was created, which was transformed into the open stock company "Aeroflot Russian International Airlines" on 28 July, 1992. At the same time, the airline began operating foreign aircraft.

25 years of civil aviation in the USSR. This poster was not created for Aeroflot's anniversary but for the anniversary of Dobrolet, an airline that started regular internal flights in the USSR in July 1923. Artist is Sergey Georgievich Sakharov. (posterplakat.com)

B-25s were used by the so-called departmental aviation. Among the owners of the B-25s there were Aeroflot, Polar Aviation, Main Directorate of Hydrometeorological Service, Main Department of Geodesy and Cartography, Ministry of Aviation Industry, Geology and Fisheries, Dalstroy Ministry of Internal Affairs and the General Directorate camps of the Mining and Metallurgical Industry of the Ministry of Internal Affairs. Also, the Gromov Flight Research Institute (known by its Soviet acronym "LII" used at least one B-25. The demilitarised B-25s were entered into the civil register and used for example as courier aircraft or as photo survey aircraft. Despite the large numbers of B-25s that the Soviet Union had, only very few have been transferred to the civil market and, unfortunately, little is known about these airplanes.

The Grazhdansky Vozdushny Flot (GVF), Civil Air Fleet, leadership counted on the rather wide application of the B-25s in civil aviation after the war. According to experts from the GVF Research Institute, the economic performance of the disarmed B-25 was approaching that of the Douglas C-47 and Lisunov Li-2. They concluded that the B-25 was the most cost-effective of military aircraft that could be used in the GVF as a cargo and cargo-passenger express. One of the Soviet civil B-25s was CCCP-A370. This was a B-25J-5 with U.S. s/n 43-27860 and was used by Aeroflot. This ship was converted to a photo survey aircraft. She had a dark camouflage scheme with light undersides and small writings on her nose. Probably, this was the standard U.S. Olive Drab/Neutral Grey camouflage scheme.

The Ministry of Aviation Industry (MAP) had a B-25 with a number starting with CCCP-I. The other digits are unknown. This airplane crashed on 28 September, 1946 during a cargo flight from Irkutsk to Ufa. The airplane ran out of fuel, crash-landed in the coniferous forest, 9 km from Chermanchet (Shitkino district of the Irkutsk region), killing four of the five crew members.

In 1955, an Olive Drab B-25D of the GVF UCT (Central Administration of the Civil Air Fleet) was used for geological research at Ukhta, the industrial city in the Komi Republic, North-Western Soviet Union. She was registered as CCCP-A1169.

A B-25 testbed with the detachable cockpit section of the DFS-346. This was a German rocket-powered swept-wing aircraft which began development during World War II. This airplane was operated by the Gromov Flight Research Institute LII. (via Carl Geust)

A civil B-25J in flight. It is hard to see, but on the rear fuselage can be read in bright letters CCCP-A1169. This ship was used for geological research at Ukhta in North-Western Soviet Union in 1955. (via Carl Geust)

CCCP-M345 was another airplane of Gidrometsluzhba. In September 1947, this B-25 was being flown from the Naberezhnaya airfield to Leningrad when an engine fire broke out after take-off. The plane crashed on a house and burned out. Only the radio operator survived from the crew. One passenger died; others were injured. A small boy was killed in the house and his mother was injured.

The airplanes of the air units of the Ministry of Geology basically carried out photography of the geological parties' work areas. In January 1950, only three airplanes were listed for this department.

In July/September 1950, the Ministry of Fisheries of the eastern regions tried to use the B-25 to search for fish accumulations in the Pacific, but unsuccessfully because they felt that the speed of the airplane was too high. In addition to one aircraft that was in the hands of Glavsakhalinrybprom (Main Administration of the Sakhalin Fishing Industry), two more B-25s were based in Moscow.

The Gidrometsluzhba (Directorate of the Hydrometeorological Service) exploited former bombers as scouts of weather and ice. Peak number of the park was in June 1947 with eight airplanes. Further, their number gradually decreased due to wear and accidents. In December 1946, a B-25 was in service with the 2nd flight of the 37th independent composite aviation detachment of the Yakutian Directorate of the Hydrometeorological Service. She had the number CCCP-M178. In September 1948, probably the same airplane was converted to a photo survey airplane for the aviation detachment of the GUGK (Yakutian Aerogeodesy Enterprise of the Main Directorate of Geodesy and Cartography).

One of the civil B-25s of the Ministry of Geology at Ukhta in 1955. This Olive Drab/Black B-25J is a civil registered airplane with black registration CCCP-A32... painted on the rear fuselage. The former nose art has apparently been overpainted. (Rusavia, via Carl Geust)

ARGENTINA

In 1944, the Polyarnaya Aviatsiya had a B-25 with the registration CCCP-N336. This was an aviation branch that was managed by the Directorate of Polar Aviation. The main task of this directorate was the development of the huge northern and eastern territories of the Soviet Union. In 1970, the Directorate was discontinued and Polyarnaya Aviatstiya was merged with Aeroflot. Polyarnaya Aviatsiya did not have much success with the polar pilots, and in January 1945, the B-25 was transferred to Dalstroy. Another B-25 of Polyarnaya Aviatsiya had the registration CCCP-N366. This airplane was used by MA-GON (Moscow Special Purposes Air Group). Later, the Chief Directorate of the Northern Sea Route, also known as Glavsevmorput (GUSMP), received another aircraft. In January 1947, she made a forced belly landing. Polar explorers repaired her and the former Dalstroy designation X-717 was replaced with H-445. The plane was used by the Moskovskaya Aviagruppa at Zakharkovo (in the territory of the present city of Khimki). This B-25 was written off in the first half of 1950.

In January 1945, a B-25 was used by NKVD-Dalstroy, probably CCCP-N336, and served as cargo ship. The Narodnyi Komissariat Vnutrennikh Del, abbreviated NKVD, (People's Commissariat for Internal Affairs) was the leading Soviet secret police organisation from 1934 to 1946. Dalstroy, also known as Far North Construction Trust, was an organisation set up in 1931 by the NKVD in order to manage road construction and the mining of gold in the Chukotka region of the Russian Far East, now known as Kolyma. Initially it was established as State Trust for Road and Industrial Construction in the Upper Kolyma Area. Dalstroy oversaw the development and mining of the area using slave labour. Over the years, Dalstroy created some 80 Gulag camps across the Kolyma region. After the 1952 reorganisation it was known as Main Directorate of Camps and Construction of the Far North. CCCP-X717 was a B-25 of MMP-Dalstroy (Moscow Special Purposes Air Group).

The planes belonging to the Ministry of Internal Affairs were just cargo aircraft. One was based at Dalstroy and two at the General Directorate of Mining and Metallurgy Camps (GULGMP) in the detachment serving the Norilsk Combine. The airplanes of the Ministry of Aviation Industry served as cargo aircraft. In September 1947, one of these (I-850) made a forced landing at Vnukovo during the flight to Kazan and crashed into a Ilyushin Il-12. The entire crew and passengers of the B-25 lost their lives.
Finally, CCCP-I850 was used as a cargo aircraft by Narkomaviaprom, or known by its initials NKAP (People's Commissariat of Aviation Industry).

One of the Latin American countries that flew with a few civil B-25s was Argentina. In 1960 and 1961, three B-25Js were supplied to Argentina from U.S. surplus sales.

LV-PWE
One airplane was a B-25J-30 with serial number 44-31318 and flew only for a few months in Argentine. She was stored at Davis Monthan AFB, Arizona, in 1958-1959 and then went to the civil market. She received the U.S. civil number N8090H. In May 1960, after she had had several owners in Miami, Florida, she was sold to Fortuna Vardago, Buenos Aires, and was registered as LV-PWE for delivery. In July 1960, she departed the U.S. for Argentina. She was then reported by error as LV-TWE and LV-TAE. In August 1960, she was modified with an overall metal nose to replace the transparent glass nose. But already in December 1960, she was sold to Raimundo Nachumow in Montevideo, Uruguay.

LV-GJX
A second airplane was registered as LV-GJX, a B-25J-30 model with the U.S. serial number 44-31498. The airplane was stored at

The B-25J-30, s/n 44-31498, registered as LV-GJX operated by Aeroexploracion SA and pictured in September 1961. Her career was comparatively short, however, and she was derelict at Aeropuerto Internacional Ministro Pistarini, known as Ezeiza International Airport by 1964 and was finally broken up in March 1968. (Hélio Hicuchi)

October 1966, this is LV-GJX in a derelict condition at Ezeiza International Airport. On the vertical stabilizer she has the number F-5 and on the nose is written, but barely visible, the name "El Cuco".
(crimso.msk.ru)

Davis Monthan after use by the USAAF. In June 1960, she went to Argentina and was acquired by Servicios Aereos Albarenque and was later operated by Aeroexploracion SA–Instituto Fototopografico Argentina, Buenos Aires. She served for two years and flew with the number F-5 and was named "El Cuco". She was withdrawn from use by 1964, was broken up and scrapped. The ship was natural aluminium finished with black engine nacelles and registration number.

LV-GXH

Airplane LV-GXH was also a B-25J-30 and had serial number 44-31173. This airplane was delivered in April 1945 to the USAAF. In December 1958, she was flown to storage and in June 1961, she was sold to Enrique Denwert in Argentina and registered as LV-GXH. From 1961 until 1971, she was used for smuggling whiskey and tobacco from Paraguay. In the 1960s, and until the middle of the 1970s, old bombers or transports of the Second World War were used for smuggling. Normally, the load came from Paraguay. The favourite provinces for smuggling were Cordoba, the north of Santa Fe, a part of Buenos Aires and Santiago del Estero. The City of Santiago del Estero is in an area close to Bolivia and Paraguay. A Governor of Santiago del Estero decided, at a certain point, that aircraft that are captured, would be seized by the province for local use as

One of the first pictures of LV-GXH in Argentina. The picture was taken in 1964 in Buenos Aires. The overall aluminium finished airplane had a white fuselage roof separated by a red stripe. The engine nacelles were black. She has the blue and white Argentine flag, without the sun, on her rudders. (Proyecto B-25 Mitchell (Huaira Bajo)

LV-GXH at Santiago del Estero in 1974. After 1976, she was left in a derelict condition at Santiago del Estero. ((Collection Nery Mendiburu)

ambulance, transport, etc. On one trip in 1969, the B-25 ran out of fuel in the province of Santiago del Estero and so she was confiscated by the province, baptized "Huaira Bajo" and flew with a provincial airline called "Empresa Provincial de Aviación Civil de Santiago del Estero" until 1976. Then she suffered a double engine failure and made an emergency landing on the runway of Santiago del Estero in the same year. Because of lack of money for an overhaul, she was pushed off the runway. For the next 35 years, she was left and exposed to the weather influences. Although she was still standing on her wheels, she showed the deterioration of many years outdoors and the actions of the curious who took souvenirs of her structure and instrumentation. In 2011, she was acquired by Proyecto B-25 Mitchell (Huaira Bajo) and is currently under restoration. The ship was overall natural aluminium finished with a white roof separated by a red stripe. The shape of this stripe varied from time to time. The engine nacelles, the registration number, and the name "Huaira Bajo" were black.

Two other pictures of the airplane around 2007-2008 before she was saved by Proyecto B-25 Mitchell (Huaira Bajo).

(Uriel Fernández, Robert Domandl)

The plane is restored in Buenos Aires by Proyecto B-25 Mitchell. During the restoration, a splendid new nose art was applied. These pictures are from October 2017 and October 2018. (Proyecto B-25 Mitchell (Huaira Bajo), Twilvina)

BOLIVIA

There is much uncertainty about the B-25s in Bolivia. Probably, a total of 13 B-25s have flown in this country, of which seven went to the Bolivian Air Force. In addition to the military aircraft of Bolivia, probably six B-25s got the civil Bolivian CP registration. These airplanes were all acquired and later sold or written off in the 1960s and 1970s. They were often used as a meat hauler.

CP-681

In the late 1960s, one B-25J was registered as CP-681. However, it is unclear whether this airplane was ever actually delivered. Further details are unknown.

CP-718

A second airplane was a B-25J-30 with s/n 44-86861. The airplane had the U.S. civil number N3482G and was exported to Bolivia in September 1964. She was registered as CP-718 and flown by Transportes Aéreos Be-

nianos SA. (TABSA). With its Head Office at La Paz, it operated cargo services between La Paz and the Departments of Beni, Santa Cruz, and Pando. It is now known as Bolivian Airways. On 2 March, 1966, the B-25 crashed at Cobija.

CP-796

Airplane CP-796 was a B-25J-20 with s/n 44-29287 and registered as N9116Z. In 1963, the airplane was converted into a tanker. On 23 August, 1966, she was delivered to Tranportes Aéreos Benianos SA for meat hauling duties. She was registered as CP-796 and crashed at Laja, Bolivia, on 18 February, 1967.

CP-808

Airplane CP-808 was also flown by Tranportes Aéreos Benianos SA. This was an ex-RCAF airplane with No. 5204 and was a B-25J-30 with s/n 44-86820. In April 1962, she was registered as N92874 and in October 1962 as N232S. In March 1967, the airplane was sold to Transportes Aéreos Benianos SA at La Paz and registered as CP-808. A month later she crashed at Itagua, Bolivia but was

CP-808 in service with Transportes Aéreos Benianos SA (TABSA) at La Paz. TABSA, was later known as Bolivian Airways. This is a B-25J-30 with U.S. serial number 44-86820. The airplane was acquired by TABSA in 1967 and was overall aluminium finished with black engine nacelles, a white fuselage top with a red dividing line in the form of a lightning flash. The red, yellow, and green striped Bolivian flag is painted on the vertical stabiliser. (Collection Wim Nijenhuis)

Right: B-25J-20, s/n 44-29287, came on the civil registry as N9116Z. In 1966, she was exported to Bolivia and registered as CP-796. This photograph shows the airplane still with her U.S. registration before delivery to Bolivia. (Collection Wim Nijenhuis)

December 1972, El Alto International Airport, La Paz, *"Bolivariana" is now painted on the nose of the B-25. In 1972, the airplane was owned by Bolivariana. This company began operating as an independent company around the year 1972. The owner of the company, Cap. Francisco García, was related to other aviation companies of the time and personally owned the ship, one of the B-25s that operated as freighters in Bolivia. (aviadejavu.ru)*

A picture of a civil B-25 at the end of her service at La Paz. The remains of the black registration number CP-915 on the fuselage and the red/yellow/green fin flash on the tail are still faintly visible. This is a B-25D-30 model with serial number 43-3308. The airplane went to Bolivia in 1970 and served with Transportes Aéreos Benianos SA. She was registered as CP-915. By 1976, she was withdrawn from use and was left derelict at La Paz. In 1987, she was shipped to the U.S.A. and restored for display as a PBJ-1D at the USMC Air-Ground Museum in Quantico, Virginia. At right, she is seen in February 1973 at El Alto Airport in La Paz. (Roy L. Stafford, Hélio Hicuchi)

repaired and in 1972 she went to Servicios Aeros Bolivianos Boliviana de Aviacon-Bolivariana, La Paz. Finally, the B-25 crashed on 21 November, 1977.

CP-915

This was a B-25 built for the RAF and seconded to the RCAF during the war. This was a B-25D-30 with s/n 43-3308. In 1964, the ship was sold to Bellomy Aviation, Miami, Florida, and registered as N8011. In 1966, she was sold to Aerovias Internacional Alianza in Panama and registered as HP-428. In April 1970, she was sold to Transportes Aéreos Benianos SA and registered as CP-915. In 1975, the ship was sold to Sudamericana, La Paz. About one year later she was withdrawn from use and was left derelict at La Paz. In 1987, she was returned to the United States by Roy M. Stafford from Jacksonville, Florida, and was restored to static airplane and displayed as USMC PBJ-1D at the USMC Air-Ground Museum in Quantico, Virginia. In 2002, the museum closed, and the B-25 was transferred to the B-25 Preservation Group. She is now on display at the Freedom Museum USA in Pampa, Texas.

CP-970

The sixth B-25 was CP-970, a B-25J-15, s/n 44-28945. In 1944, this was General Henry H. "Hap" Arnold's private transport airplane. The B-25 was selected for the conversion, duplicating almost bolt-for-bolt the modifications that had been incorporated in General Eisenhower's B-25J. The nose was fully faired over and a series of antennas were mounted underneath the forward fuselage.

A direction finder loop was installed on top of the nose. In 1960, after storage at Davis Monthan, the airplane was sold to Edwards Petroleum Co., Fort Worth, Texas, and registered as N3184G. One of the following owners was the Bendix Corporation, Baltimore, Maryland. In 1972, the airplane was acquired by Transportes Aéreos Benianos, La Paz and was registered as CP-970. The ship crashed on 7 June, 1976.

Post-war picture of number 44-28945 still in service with the USAF. She had a white fuselage top and the Bolling Field insignia painted on the nose. The airplane was converted to a VIP transport. Mid-1970's, the airplane flew in Bolivia with the registration CP-970. (Collection Wim Nijenhuis)

An extremely rare picture of CC–CAK. This was ex N3506G, ex N175LT and ex CC–CLG of Línea Aérea Taxpa. (Tony Sapienza)

Very rare pictures of CC–CLK of Pacific Air Cargo Air Line in 1960. (via Claudio Cáceres Godoy)

CHILE

Apparently, Chile had two B-25Js with a civil Chilean registration. Unfortunately, very little is known about these airplanes. One airplane was a B-25J-25, s/n 44-30483, and was stored at Davis Monthan in 1958-1959. In August 1959, she was sold to Ace Smelting Inc., Phoenix, Arizona, respectively Charlene J. Williams, Los Angeles, California, and registered as N3504G. In March 1960, she was sold to Gaston Neito/Pacific Air Cargo, Santiago, Chile, and registered as CC-CLK. The B-25 carried several air loads in Chile, including during humanitarian aid flights following the May 1960 earthquake. By November 1960, the airplane had already been withdrawn from use and was stored at Santiago-Los Cerillos. She was removed from Chilian civil registry in February 1963. The other airplane was a B-25J-30 with s/n 44-31487. After she had been stored at Davis Monthan, she got a private owner and was registered as N3506G and later as N175LT. In September 1961, she was sold to Gaston Neito/Pacific Air Cargo, Santiago, and registered as CC-CLG.

After a time of operations, the airline falls into insolvency, and the plane went up for auction in June 1963, being acquired by Oscar Squella Avendaño from Taxpa Airline, she was re-registered as CC-CAK. However, it should be noted that some sources mistakenly mention that CC-CLG and CC-CAK were two different airplanes. In January 1964, the B-25 was reported crashed in Paraguay. Línea Aérea Taxpa, in the brand name Taxpa Chile, was a Chilean airline that had started flying in 1958 and operated with various airplanes until 1979. The company offered air taxi services, pest control flights, aerial survey, and rescue services, but also flew regular service to Robinson Crusoe Island in the Pacific Ocean during the season.

COSTA RICA

Like Chile, Costa Rica seems to have had two B-25s with a civil registration. But also, of these airplanes very little is known. One airplane was briefly registered as TI-1046C and seems to have served with Expreso Aereo Costarricenses SA. This was a Costa Rican operator, known as LADECA, formed in April 1951 to undertake air taxi and charter work and crop spraying from San Jose, as well as domestic cargo charters. It operated in the 1950s and 1960s.

A second B-25 was B-25J-30, s/n 44-86834. After her storage at Davis Monthan AFB, she was sold to National Metals Inc., Phoenix, Arizona, and registered as N9495Z. In

A colourful B–25 with the Mexican registration XB–MOP. Apparently, this airplane flew for a short time with the Costa Rican registration number TI–1026L. (Gary Kuhn)

The civil Mexican B–25H–5, in 1955 registered as XC–BIV with the name "Cuauhtémoc". He was the last Aztec Emperor, ruling from 1520 to 1521. He was tortured and killed by the Spanish conquistador Hernán Cortés in 1525. The airplane is known for having served the President of Mexico and his staff. It was modified circa 1956 and fitted with a hydraulic vault lift to load gold bullion and with 12 passenger seats and two sleeping berths. The Mexican flag is just visible on her vertical tail surface. (Collection Wim Nijenhuis, Gary Kuhn)

Most likely, XB-DAD was a B-25J-20 with U.S. serial number 44-29724. In 1958, she was registered as N7706C. In August 1958, she was sold to Mike Casali & Carlos Wong Medina, Mexico and registered as XB-DAD. In 1968, she went to Turbo Mex, Mexico City and was struck-off charge in April 1968. (aviadejavu.ru, Brian Baker via Gary Kuhn)

July 1960, she went to Mar-Tod Exports Co., Corpus Christi, Texas. Thereafter, she was registered with the Costa Rican number TI-1026L. Unfortunately, there are no more details known of this Costa Rican service. Mid-1960s, she made a forced landing at Campeche, Mexico, and was confiscated by the Mexican Government. In August 1966, she was registered as XB-MOP. She was stored damaged and derelict at Miami, Florida in 1968.

MEXICO

After the Second World War, Mexico had several civil B-25s during the 1960s and 1970s. They were nearly all B-25J models. The first, a B-25H-5, s/n 43-4645, had the U.S. serial number N123A. In 1955, she was exported to Mexico and registered as XC-BIV to Banco de México S.A. and was operated for the Mexican President and his staff. Four other airplanes were registered as XB-DAD, XB-GAR, XB-HEY and XB-MOP. They

This is s/n 44-30323, a B-25J-25. In 1958, after storage at Davis Monthan, she was sold and registered as N9656C. In June 1959, she was sold to Servicios Aéreos de América SA, Mexico City and registered as XB-GAR. In 1969, she was sold to The Diners Club de Mexico, Mexico City. She was removed from the civil registry in September 1969. (Gary Kuhn)

This is XB-GAR in a later colour scheme, at right with a severely damaged nose. (America Vuela, aviadejavu.ru)

Another civil B-25 was XB-HEY. These pictures were taken in the early 1960s. She is a B-25J-35 with s/n 45-8843. She was purchased by Tallmantz Aviation and used in the film "Catch-22". The B-25 did not fly for the film but was used for a sequence depicting a B-25 crash. The burnt remains of the airplane were buried on the airstrip site at San Carlos when the film was completed. (Brian Baker via Gary Kuhn, Collection Ed Coates)

were used by civil operators and XB-HEY was used in the film "Catch-22". This film was made in 1969 and is a satirical war film adapted from the book of the same name by Joseph Heller. In the famous opening scene of this film, sixteen B-25s taxi out to the runway in Mexico and make a mass take-off. XB-HEY was one the B-25s used in this film. She was a B-25J-35, s/n 45-8843. In April 1959, after storage at Davis Monthan, she was sold to Edward Tabor, Los Angeles, California, and registered as N8091H. She was exported to Mexico, date unknown, and became XB-HEY. She was located at an airfield near Guyamas, Mexico, and her last listed Mexican owner was Charles E. Rector of Guaymas, Sonora. In January 1969, she was purchased by Tallmantz Aviation at Loreto, Mexico, for using in the film "Catch-22", see chapter Tallmantz. She did not fly for the film; she was used for a sequence depicting a B-25 crash. The burnt remains of the airplane were buried on the airstrip site at San Carlos when the film was completed.

XB-MOP was a B-25J-30 with s/n 44-86834. Mid-1960s, she made a forced landing at Campeche, Mexico, and was confiscated by the Mexican Government. In August 1966, she was registered as XB-MOP. This airplane also flew with the Costa Rican number TI-1026L, further details unknown.

XB–HEY is seen here in January 1969 at Loreto, Mexico. The airplane was prepared for its ferry flight from Loreto to the "Catch–22" airstrip on the other side of the Sea of Cortez. (Collection Charles Rector)

OTHER CIVIL MEXICAN B-25S

Three other B-25s were in Mexico, but they had no civil registrations. Two B-25s were displayed in the Bosque de San Juan de Aragón and one was displayed in the Bosque de Chapultepec. Both are large public parks in Mexico City. The U.S. registered N9877C and N9623C were placed in the park San Juan de Aragón. As civil aircraft, both were together since 1965 when they were offered for sale by Allied Aircraft Sales in Phoenix. The same year, both were purchased by the International Civil Aviation Organisation and sent to Civil Aviation Training Centre in Mexico City. Then donated to the Mexico City's government in 1985 and placed on display in the park. Unfortunately, both airplanes soon became a wreck. They were there in poor condition, both frames had empty cockpits and were covered inside and outside with graffiti. In 2011, the airplanes were removed from the park and rebuilt to a static display with Mexican Air Force markings.

The third B-25, ex U.S. N92872, was placed in the Chapultepec Park and is still there on a pole.

Above: *Another B-25 in the Bosque de San Juan de Aragón was N9623C, a B-25J-25 with s/n 44-30692. In 1990, she was displayed in the park. Like N9877C, the airplane was in a poor condition (Booxmiis)*

Airplane N9877C was a B-25J-20, s/n 44-29145, and registered in 1958 after storage at Davis Monthan, Arizona. In 1965, she went to Centro Internacional de Adiestramiento en Aviación Civil at Mexico City. She was donated to the Departamento del Distrito Federal circa 1985 and placed on display in the Bosque de San Juan de Aragón. This is a public park located in the Delegation Gustavo A. Madero, near the International Airport of Mexico City. The airplane never got a Mexican registration, but a code XL-7 was added as well as a shark mouth motif. Unfortunately, the airplane became a wreck, with the rear fuselage and tail section missing. (aviadejavu.ru, Booxmiis)

Both airplanes were removed from the San Juan de Aragón Park in 2011. The fuselage of N9623C and parts from both B-25s were used to rebuild to static display as BMM-3503 of the Mexican Air Force. The work was done by the 4th Maintenance Echelon of the Mexican Air Force at Santa Lucia Air Base. This is the rebuilt airplane around 2015. Note the remarkable nose front pieces and top turret. *(Booxmiis)*

This is N92872, a B-25J-20 with s/n 44-29128, at the Museo Tecnológico de la Comisión Federal de Electricidad in Mexico City. This airplane went to the museum in 1972 and was put on a pole in Chapultepec Park.
Left: In the colours from 1975 and in fake Mexican Air Force markings around 2015.
(Jun Oizumi, PhilG8)

Bottom: N9958F in the 1960s in California. She was painted in two different colour schemes. At right, an attractive colour scheme with a partially blue fuselage top and black with yellow bordered arrow. In the 1970s, this airplane was displayed in Mexico and registered as XB-DOF.
(Collection Wim Nijenhuis, Nick Williams)

Another B-25 with a Mexican registration was XB-DOF, but further details of the Mexican service are unknown. This was a B-25-J-20, s/n 44-29678, an ex-RAF ship and after the war in service with the French Groupe de Transport et de Liaisons Aériennes I/60 GLAM. The airplane crashed on 28 July, 1951 at Luxeuil, France, and was summarily repaired and returned to the U.S. In 1953, she was registered as N9958F. Apparently, she was exported to Mexico in the 1970s and registered as XB-DOF. She was displayed in Mexico City and scrapped in 1989.

N2825B in 1977. She was assigned the registration XB-GOG, but never flew in Mexican service. *(Jeff Billings)*

The TB–25N with s/n 44-29123 was sold to Banco de México. In 1958, she became the U.S. registration N9869C. *(Collection Wim Nijenhuis)*

Three other Mexican registrations are mentioned by several sources, XB-GOG, XB-NAJ and XB-HAZ. Airplane XB-GOG was a B-25 with s/n 40-2168, one of the very early produced B-25s. In January 1947, after disposal at McClellan Field, California, she was sold to Charles Bates of Chattanooga, Tennessee and a ferry permit was issued under an assigned Mexican registration of XB-GOG. However, the aircraft was kept in the United States and assigned the registration NL75831 in April of that same year. In 1948, she was sold to Bankers Life and Casualty Company from Chicago, Illinois. Later, she received the number N2825B. So, the airplane never flew in Mexican service and is nowadays still flying in the U.S. as "Miss Hap". XB-NAJ was a B-25J-10 with s/n 43-36074. She had the U.S. civil number N9080Z. Regrettably, of the other aircraft XB-HAZ, nothing is known.

Finally, the B-25 trainer with s/n 44-29123 was sold to the Bank of Mexico. This airplane

was stored for disposal at Davis Monthan in December 1957. She was sold to G.E. Zarski, Brownsville, Texas in September 1958 and registered as N9869C. The next month, she was sold to Banco de México at Mexico City, but further details are unknown.

PANAMA

Panama had only one civil B-25. She was a B-25D-30 with USAAF s/n 43-3308 and this was the same airplane as mentioned earlier under the Bolivian Mitchells. She was built for the RAF as a Mitchell Mk. II and seconded to the Royal Canadian Air Force during the war as KL156. She continued in service with the RCAF until November 1961. In 1964, she was sold to Bellomy Aviation, Miami, Florida, and registered as N8011. In 1966, she was exported to Panama to Aerovias Internacional Alianza in Panama and registered as HP-428. In April 1970, about

four years later, she was reported as sold to Transportes Aéreos Benianos of La Paz, Bolivia. There, her new registration was CP-915. By 1976, after several years in Bolivian service, the airplane was reported derelict in La Paz. Finally, she was returned to the United States in 1987 and is nowadays on display at the Freedom Museum USA in Pampa, Texas.

PARAGUAY

As far as known, there was one B-25 with a civil Paraguayan registration. This was ZP-BDF, but nothing else is known about this airplane. In the years 1960 to 1974, three other B-25s were noted impounded at Paraguay's main airport at Asunción, nowadays called Aeropuerto Internacional Silvio Pettirossi, Asunción. By 1980, they were all three noted derelicts. Despite being in Paraguay, these B-25s had U.S. registrations. They were N8193H (B-25J-30, s/n 44-30976), N8194H (B-25J-30, s/n 44-86749) and N9076Z (B-25J-25, s/n 44-30772).

This is the B-25 HP-428 in 1969, at the Aeropuerto Internacional de Tocumen, the international airport of Panama City. In 1966, she was exported to Panama to Aerovias Internacional Alianza. The airplane is natural aluminium finished with a white fuselage top and a yellow lightning flash on the fuselage and on the engine nacelles. Note the Panamian flag on the rear fuselage. (Gary Kuhn)

Hard to see, but this is ZP-BDF photographed at Asunción Airport. (Tony Sapienza)

A rare picture from August 1968 taken at Guarani International Airport in Minga Guazú, a city of Gran Ciudad del Este in the Alto Paraná Department of Paraguay. Next to the Curtiss Commando is one of the Mitchells in Paraguay. (Tony Sapienza)

This is N9076Z in May 1965 at Salto, Uruguay. At the time, she was owned by companies in Miami and probably used as smuggler. Later, she was impounded at Asunción. (Gary Kuhn)

VENEZUELA

As far as known, only one B-25 flew in Venezuela in the 1960s and 1970s with civil markings. This was a B-25J-30 with the U.S. serial number 44-86773. She was delivered to the RCAF with the serial number 5282. After storage at Claresholm, Alberta, she was sold to Cascade Drilling Company of Calgary, Alberta. End of 1963, she was sold to Bellomy Aviation of Miami, Florida, and registered as N8010. Later sold to Jovan Corporation of Florida. In December 1964, she went to Venezuela for service with Tranarg C.A., an aerial photography company at Caracas. She was registered as YV-E-IPU, but ca.1975 also reported as YV-19CP when in service with SAPROLATE. The airplane was damaged in an accident on 11 August, 1977. In 1983, she was reported derelict at Caraca-Maiquetia, Venezuela.

The only known civil Venezuelan B-25. This is the ex USAAF B-25J-30 s/n 44–86773. In December 1964, after use by the Royal Canadian Air Force, she went to Tranarg C.A. at Caracas and registered as YV-E-IPU. She was used by Tranarg in aero photographical surveys. Later, the airplane was in service with SA Proyectos Y Levantamientos Aéreos Y Terrestres (SAPROLATE). (Boudewijn Deurvorst)

The airplane at Caracas in November 1971. The airplane of the company Tranarg was white with a black fuselage bottom and engines. The Venezuelan flag is painted on her rudders. (Collection Wim Nijenhuis)

FINAL WORD

Currently, there are approximately 120 surviving B-25s scattered all over the world, but mainly in the United States. Most of these are on static display in museums, and about 40 are still airworthy. Of all these airplanes, most are J-models with the post-war engine modifications of the fifties. Finally, there are still approximately 20 B-25 remnants in various museums or stored at companies.

The B-25 was a good airplane. But that there are still so many extant, more than 75 years after its first production, is surprising. Thanks to ambitious warbird enthusiasts we can still enjoy them. It is hoped that this will persist and that the number of planes will not reduce. Perhaps through restoration or new finds, the number might even increase. In any case, it is a tribute to all those who at Inglewood and Kansas City have given their heart and soul to design and to produce this amazing aircraft.

In the following Appendices 1 to 3, an overview is given of all the B-25s mentioned and described in this book, with their civil registration numbers and USAAF serial numbers as well as an overview of the airworthy airplanes and the various owners and museums.

Appendix 1
Civil registration numbers with corresponding USAAF numbers

U.S. civil reg.	USAAF serial nr.	Model	Foreign countries civil reg.
N10V, NL90339	43-4432	B-25H-5	
N25NA, N92889	44-86725	B-25J-30	
N32T, NL75754	45-8882	B-25J-35	
N123A	43-4645	B-25H-5	XC-BIV (Mexico)
N201L, N5249V	44-30606	B-25J-25	
N543VT, N325N	44-86698	B-25J-30	C-GUNO, CF-NWU (Canada)
N898BW, N3681G	45-8898	B-25J-35	
N1042B	44-30823	B-25J-25	
N1203, NX1203	43-4643	B-25H-5	
N1582V, N37L	43-4899	B-25H-10	
N2825B, NL75831, N75831	40-2168	B-25	XB-GOG (Mexico)

U.S. civil reg.	USAAF serial nr.	Model	Foreign countries civil reg.
N2835G, N58TA	44-29910	B-25J-20	
N2849G	44-30077	B-25J-25	
N2854G	44-29812	B-25J-20	
N2887G	44-86716.	B-25J-30	
N2888G	44-86872	B-25J-30	
N3155G	44-30832	B-25J-25	
N3156G, N5833B	45-8884	B-25J-35	C-GCWJ (Canada)
N3160G, N27493	44-29869	B-25J-20	
N3161G	44-30324	B-25J-25	
N3174G	44-31032	B-25J-30	
N3184G	44-28945	B-25J-15	CP-970 (Bolivia)
N3337G	44-86891	B-25J-30	

U.S. civil reg.	USAAF serial nr.	Model	Foreign countries civil reg.
N3339G	43-4030	B-25J-1	
N3398G	44-30761	B-25J-25	
N3438G	44-86797	B-25J-30	
N3441G	44-28738	B-25J-15	
N3442G	44-86715	B-25J-30	
N3443G	44-30470	B-25J-25	
N3446G	44-31466	B-25J-30	
N3453G	44-86844	B-25J-30	
N3476G	44-28932.	B-25J-15	
N3481G, N345TH	44-31385	B-25J-30	
N3482G	44-86861	B-25J-30	CP-718 (Bolivia)
N3503G	44-30085	B-25J-25	
N3504G	44-30483	B-25J-25	CC-CLK (Chile)
N3506G, N175LT	44-31487	B-25J-30	CC-CLG, CC-CAK (Chile)
N3507G	44-86843	B-25J-30	
N3512G, N43BA, N747AF	44-30456	B-25J-25	C-GTTS (Canada)
N3513G	44-31489	B-25J-30	
N3514G	44-86786	B-25J-30	
N3515G	44-31042	B-25J-30	
N3516G, N61821	44-29035	B-25J-15	
N3521G	44-86853	B-25J-30	
N3523G, N25GL	44-29465	B-25J-20	
N3525G	44-86694	B-25J-30	
N3675G	44-30423	B-25J-25	
N3676G	44-29835	B-25J-20	
N3677G	44-86782	B-25J-30	
N3680G	45-8887	B-25J-35	
N3682G	44-86708	B-25J-30	
N3695G	44-28926	B-25J-15	
N3698G	44-29507	B-25J-20	N320SQ, PH-XXV (Netherlands)
N3699G, N30801	44-30801	B-25J-25	
N3774	43-3634	B-25D-35	CF-NWV (Canada)
N3968C, NL75635,	41-13251	B-25C-1	
N3969C	43-36075	B-25J-10	
N3970C	43-4999	B-25H-10	
N5078N, N122B, N2DD, N2XD	41-29784	B-25D	
N5126N	44-30975	B-25J-30	
N5239V	44-30079	B-25J-25	
N5256V	43-28222	B-25J-10	
N5262V	44-86785	B-25J-30	
N5277V, N225AJ, N744CG	44-28866	B-25J-15	CF-OND (Canada)
N5455V	44-30721	B-25J-25	
N5548N	43-4106	B-25H-1	
N5672V, N69345	45-8835	B-25J-35	CF-DKU, C-FDKU (Canada)

U.S. civil reg.	USAAF serial nr.	Model	Foreign countries civil reg.
N5857V, NL5857V	44-30982	B-25J-30	
N5865V	44-30988	B-25J-30	
N6116X	44-86791	B-25J-30	VH-XXV (Australia)
N6123C	44-86893	B-25J-35	
N6578D	44-31508	B-25J-30	
N7493C	43-3376	B-25D-30	
N7614C	44-31171	B-25J-30	
N7674	44-30421	B-25J-25	CF-OVN (Canada)
N7681C	44-86701	B-25J-30	
N7687C	44-28925	B-25J-15	
N7693, N198W	44-30937	B-25J-30	
N7706C	44-29724	B-25J-20	XB-DAD (Mexico)
N7707C	44-30690	B-25J-25	
N7946C	44-28938	B-25J-15	
N7947C	44-30129	B-25J-25	
N8010	44-86773	B-25J-30	YV-E-IPU, YV-19CP (Venezuela)
N8011	43-3308	B-25D-30	CP-915 (Boliva), HP-428 (Panama)
N8013	44-31491	B-25J-30	
N8090H	44-31318	B-25J-30	LV-PWE (Argentina)
N8091H	45-8843	B-25J-35	XB-HEY (Mexico)
N8093H, N190V	45-8851	B-25J-35	
N8163H	44-86747	B-25J-30	
N8193H	44-30976	B-25J-30	
N8194H	44-86749	B-25J-30	
N8195H	44-30748	B-25J-25	
N8196H, N6116X	44-86791	B-25J-30	VH-XXV (Australia)
N9075Z	45-8896	B-25J-35	
N9076Z	44-30772	B-25J-25	
N9077Z, N25YR	43-27868	B-25J-5	
N9079Z	44-30734	B-25J-25	
N9080Z	43-36074	B-25J-10	XB-NAJ (Mexico)
N9086Z	44-30613	B-25J-25	
N9088Z	44-30733	B-25J-25	
N9089Z	44-30861	B-25J-25	G-BKXW (United Kingdom)
N9090Z, N600DM, N333RW	44-86734	B-25J-30	
N9091Z	44-86800	B-25J-30	
N9115Z	44-29366	B-25J-20	
N9116Z	44-29287	B-25J-20	CP-796 (Bolivia)
N9117Z	44-29199	B-25J-20	
N9167Z, N345BG	44-86777	B-25J-30	
N9170Z	44-30187	B-25J-25	
N9333Z	44-86772	B-25J-30	
N9443Z	44-28765	B-25J-15	
N9444Z, N943	44-29943	B-25J-25	
N9446Z	44-30737	B-25J-25	

U.S.	USAAF	Model	Foreign countries
civil reg.	serial nr.		civil reg.
N9451Z	44-30493	B-25J-25	
N9452Z	44-30649	B-25J-25	
N9455Z	44-30210	B-25J-25	
N9456Z	44-29939	B-25J-25	
N9462Z	44-30535	B-25J-25	
N9463Z	44-31004	B-25J-30	
N9494Z	44-30925	B-25J-30	G-BWGR (United Kingdom)
N9495Z	44-86834	B-25J-30	TI-1026L (Costa Rica), XB-MOP (Mexico)
N9552Z, N125AZ	43-35972	B-25J-10	
N9582Z	44-30607	B-25J-25	
N9610C	44-30585	B-25J-25	
N9613C	44-30377	B-25J-25	
N9621C	45-8811	B-25J-35	F-AZID, F-AZZU (France), HB-RDE (Switzerland)
N9622C, N17666	44-30243	B-25J-25	
N9623C	44-30692	B-25J-25	
N9637C	44-29249	B-25J-20	
N9639C, N87Z	44-86873	B-25J-30	
N9641C	44-30010	B-25J25	
N9642C	44-29725	B-25J-20	
N9643C	44-86758	B-25J-30	
N9655C	44-30159	B-25J-25	
N9656C	44-30323	B-25J-25	XB-GAR (Mexico)
N9753Z	44-31504	B-25J-30	
N9754Z	44-30478	B-25J-25	
N9856C	43-28204	B-25J-10	
N9857C, N1943J, N26795	43-28059	B-25J-5	C-GTTM (Canada)
N9865C	44-28834	B-25J-15	
N9866C	44-28833	B-25J-15	
N9868C	44-29919	B-25J-25	
N9869C	44-29123	B-25J-20	
N9877C	44-29145	B-25J-20	XL-7 (Mexico)
N9899C	44-29127	B-25J-20	
N9936Z	44-30756	B-25J-25	
N9937Z	43-3910	B-25J-1	
N9958F	44-29678	B-25J-20	XB-DOF (Mexico)
N9991Z	44-30996	B-25J-30	
N10564	44-29887	B-25J-20	
N41123	44-30254	B-25J-25	CF-MWC (Canada)
N52998, N102J	44-31280	B-25J-30	
N62163	44-86697	B-25J-30	
N67998, N96GC	43-4336	B-25H-1	
N75150	45-8824	B-25J-35	
N75755, NL75755	45-8883	B-25J-35	C-GCWM, C-GCWA (Canada)
N86427	44-29121	B-25J-15	

U.S.	USAAF	Model	Foreign countries
civil reg.	serial nr.		civil reg.
N88972	43-3318	B-25D-30	CF-OGQ (Canada), G-BYDR (United Kingdom)
N92872	44-29128	B-25J-15	
N92874, N232S	44-86820	B-25J-30	CP-808 (Bolivia)
N92875	44-86727	B-25J-30	
N92880	44-30947	B-25J-30	CF-NTP (Canada)
NL2424, NX2424	45-8830	B-25J-35	
NL66548	45-8829	B-25J-35	
	43-27860	B-25J-5	CCCP-A370 (Soviet Union)
	44-28898	B-25J-15	
	44-30641	B-25J-25	CF-NTS (Canada)
	44-30791	B-25J-25	
	44-31173	B-25J-30	LV-GXH (Argentina)
	44-31399	B-25J-30	CF-NTV (Canada)
	44-31493	B-25J-30	CF-NTX (Canada)
	44-31498	B-25J-30	LV-GJX (Argentina)
	44-86724	B-25J-30	CF-NTU (Canada)
	44-86726	B-25J-30	
	44-86728	B-25J-30	CF-NTW (Canada)
			CCCP-1 (Soviet Union)
			CCCP-A1169 (Soviet Union)
			CCCP-M178 (Soviet Union)
			CCCP-M345 (Soviet Union)
			CCCP-A32 (Soviet Union)
			CCCP-N336 (Soviet Union)
			CCCP-N366 (Soviet Union)
			X-717, H-445 (Soviet Union)
			I-850 (Soviet Union)
			CP-681 (Bolivia)
			TI-1046C (Costa Rica)
			XB-HAZ (Mexico)
			ZP-BDF (Paraguay)

Appendix 2
Owners and museums with civil B-25s

U.S.A.
Firefighting and agricultural duties:
- Abe's Aerial Service Inc., Safford, Arizona
- Aero Flight Inc., Troutdale, Oregon
- Aero Insect Control Inc., Rio Grande, New Jersey
- Aero Retardant Inc., Fairbanks, Alaska
- Aero Spray Inc., Vancouver, Washington
- Aero Union Corp., Chico, California
- Angel's Aerial Service, Pico-Rivera California
- Aviation Specialties Inc., Mesa, Arizona
- Biegert Brothers Inc., Lincoln, Nebraska / Phoenix, Arizona
- Blue Mountain Air Service / Hillcrest Aircraft Co. Inc., La Grande, Oregon
- Cal-Nat Airways, Grass Valley, California
- Christler and Avery Aviation Co., Greybull, Wyoming
- CISCO Aircraft, Lancaster, California
- Colco Aviation Inc., Anchorage, Alaska
- Crowl Dusters, Phoenix, Arizona
- Donaire Inc., Phoenix, Arizona
- Dothan Aviation Corp., Dothan, Alabama
- Frontier Flying Service, Fairbanks, Alaska
- Idaho Aircraft Company, Boise, Idaho
- Johnson Flying Service, Missoula, Montana
- Parsons Airpark Inc., Carpinteria, California
- Paul Mantz Air Services, Santa Ana, California
- Red Dodge Aviation Inc., Anchorage, Alaska
- L.K. Roser, Phoenix, Arizona
- Sonora Flying Service, Columbia, California
- Sprayair Ltd., Scramento, California
- Sprung Aviation, Tucson, Arizona
- Edgar L. Thorsrud, Missoula, Montana
- Trans-West Air Service / Aerial Applicators Inc., Salt Lake City, Utah
- E.D. Weiner, Los Angeles, California
- Wenairco Inc., Wenatchee, Washington

Maintenance, restoration, and trading:
- Aero Trader, Chino, California
- Grand Central Aircraft Company, Glendale, California,
- Hayes Aircraft Corporation, Birmingham, Alabama
- Hughes Aircraft Company, Culver City, California
- L.B. Smith Aircraft Corporation, Miami, Florida
- John J. Stokes, San Marcos, Texas
- Tom Reilly, Kissimmee, Florida/Douglas, Georgia

Various companies/owners:
- Aerocrafters Inc., Santa Monica, California
- Albert Trostel & Sons Company, Milwaukee, Wisconsin
- Arthur Jones, Slidell, Louisiana
- Avirex, Tenafly, New Jersey
- Bendix Aviation Corporation, Los Angeles, California
- Claire Aviation Inc., Philadelphia, Pennsylvania
- Doris Duke, New York, New York
- Husky Oil Company, Cody, Wyoming
- Inter American Minerals Corp., New York
- John Ward Aviation, Coulterville, California
- National Motor Bearing Co., Redwood City, California
- North American Aviation, Inglewood, California
- Northern Pump Company, Minneapolis, Minnesota
- Oil Tool Corp., Long Beach, California
- Radio Station KOB, Albuquerque, New Mexico
- D. Richard Lambert, Plainfield, Illinois
- Russ & Don Newman / Old Glory Inc., Tulsa, Oklahoma
- San Fernando Drag Strip, San Fernando, California
- Scott Aero Service Inc., Long Beach, California
- Shell Aviation Company, New York
- Southwest Aviation, Las Cruces, New Mexico
- Television Associates, Michigan City, Indiana
- Timken Roller Bearing Company, Canton, Ohio
- Tripoints Associates, Miami, Florida
- Walter Soplata, Newbury, Ohio
- Washington Wilbert Vault Works, Laurel, Maryland

Museums:
- Air Zoo, Kalamazoo, Michigan
- American Aeronautical Foundation, Camarillo, California
- American Airpower Museum, Farmingdale, New York
- Big Kahuna's Water and Adventure Park, Destin, Florida
- Cavanaugh Flight Museum, Addison, Texas
- Champaign Aviation Museum, Urbana, Ohio
- Collings Foundation, Stow, Massachusetts
- Commemorative Air Force, Dallas, Texas
- Delaware Aviation Museum Foundation, Georgetown, Delaware
- EAA Air Museum, Oshkosh, Wisconsin
- Erickson Aircraft Collection, Madras, Oregon
- Fagen Fighters WWII Museum, Granite Falls, Minnesota
- Fargo Air Museum, Fargo, North Dakota
- Flying Heritage & Combat Armor Museum, Everett, Washington
- Flying Tiger Air Museum, Paris, Texas
- Freedom Museum USA, Pampa, Texas
- Historic Flight Foundation, Paine Field, Washington
- Kansas City Warbirds Inc., Kansas City, Missouri
- Lauridsen Aviation Museum, Buckeye, Arizona
- Lewis Air Legends, San Antonio, Texas
- Liberty Aviation Museum, Port Clinton, Ohio
- Lone Star Flight Museum, Houston, Texas
- Lyon Air Museum, Santa Ana, California
- March Field Air Museum, Riverside, California
- Mid-America Air Museum, Liberal, Kansas

- Mid America Flight Museum, Mount Pleasant, Texas
- Mid-Atlantic Air Museum, Reading, Pennsylvania
- Milestones of Flight Museum, Lancaster, California
- Military Aviation Museum, Virginia Beach, Virginia
- Movieland of the Air Museum, Santa Ana, California
- National Museum of World War II Aviation, Colorado Springs, Colorado
- New England Air Museum, Windsor Locks, Connecticut
- Palm Springs Air Museum, Palm Springs, California
- Paul Bunyan Land, Brainerd, Minnesota
- Planes of Fame Air Museum, Chino, California
- Polar Aviation Museum, Anoka County, Minneapolis
- SST Aviation Museum, Kissimmee, Florida
- Strategic Air Command & Aerospace Museum, Ashland, Nebraska
- Texas Flying Legends Museum, Houston, Texas
- The Oklahoma Museum of Flying, Bethany, Oklahoma
- Tri-State Warbird Museum, Batavia, Ohio
- Victory Air Museum, Mundelein, Illinois
- Warbirds of Glory Museum, Brighton, Michigan
- Wiley Sanders Warbird Collection, Troy, Alabama
- Wings of History Museum, San Martin, California
- Yanks Air Museum, Chino, California
- Yankee Air Museum, Ypsilanti, Michigan

U.S. military bases:
- Castle Air Museum, Atwater, California
- Grand Forks AFB, North Dakota
- Grissom Air Museum, Peru, Indiana
- Hill Aerospace Museum, Ogden, Utah
- Hurlburt Field Memorial Air Park, Okaloosa County, Florida
- Malmstrom AFB Museum, Great Falls, Montana
- Maxwell AFB, Montgomery, Alabama
- Museum of Aviation History, Warner-Robins, Georgia
- National Naval Aviation Museum, Pensacola, Florida
- Pacific Aviation Museum Pearl Harbor, Honolulu, Hawaii
- The Flying Leatherneck Aviation Museum, San Diego, California
- USS Alabama Battleship Memorial Park, Mobile, Alabama

American film industry:
- Filmways, Inc., Hollywood, California
- Tallmantz Aviation Inc., Santa Ana, California
- Steve Hinton, Chino, California

Canada:
- Aurora Aviation Ltd., Edmonton, Alberta
- Avaco Services, Salmon Arm, British Columbia
- G&M Aircraft Ltd., St. Albert, Alberta
- Hicks and Lawrence Ltd., Ostrander, Ontario
- Holden Aviation Services, Ltd., Lamont, Alberta
- Northwestern Air Lease Ltd., St. Albert, Alberta
- Robert Diemert, Carman, Manitoba

- Alberta Aviation Museum, Edmonton, Alberta
- Canadian Warplane Heritage Museum, Hamilton, Ontario
- Reynolds-Alberta Museum, Wetaskiwin, Alberta

United Kingdom:
- John Hawke
- The American Air Museum, Duxford
- Lincolnshire Aviation Heritage Centre, East Kirkby
- Aces High Ltd.
- Southend Historic Aircraft Museum
- The Fighter Collection

Civil B-25s in other countries:
- The Netherlands
- Belgium
- France
- Switzerland
- Austria
- Spain
- Australia
- Soviet Union
- Dobrolet / Aeroflot
- Argentina
- Bolivia
- Chile
- Costa Rica
- Mexico
- Panama
- Paraguay
- Venezuela

Appendix 3
Airworthy B-25 Mitchells

Aircraft name	USAAF serial nr.	Civil reg.nr.	Model	Owner	Location
Axis Nightmare	45-8898	N898BW	B-25J	Tri-State Warbird Museum	Batavia, Ohio
Barbie III	43-4106	N5548N	B-25H	Cavanaugh Flight Museum	Addison, Texas
Berlin Express	43-4432	N10V	B-25H	EAA Aviation Museum	Oshkosh, Wisconsin
Betty's Dream	45-8835	N5672V	B-25J	Texas Flying Legends Museum	Houston, Texas
Briefing Time	44-29939	N9456Z	B-25J	Mid-Atlantic Air Museum	Reading, Pennsylvania

Aircraft name	USAAF serial nr.	Civil reg.nr.	Model	Owner	Location
Bridge Buster	44-30254	N41123	B-25J	Flying Heritage & Combat Armor Museum	Everett, Washington
Champaign Gal	44-28866	N744CG	B-25J	Champaign Aviation Museum	Urbana, Ohio
Devil Dog	44-86758	N9643C	B-25J	Comm. Air Force, Devil Dog Squadron	Georgetown, Texas
Doolittle Raiders Special Delivery	44-86734	N333RW	B-25J	Lone Star Flight Museum	Houston, Texas
Executive Sweet	44-30801	N30801	B-25J	American Aeronautical Foundation	Camarillo, California
Georgia Mae	44-86785	N5262V	B-25J	Wiley Sanders Warbird Collection	Troy, Alabama
Georgie's Gal	44-86777	N345BG	B-25J	Liberty Aviation Museum	Port Clinton, Ohio
God and Country	44-30823	N1042B	B-25J	Mid-America Flight Museum	Mount Pleasant, Texas
Grumpy	43-3318	N88972	B-25D	Historic Flight Foundation	Paine Field, Washington
Guardian of Freedom	44-29465	N25GL	B-25J	Lyon Air Museum	Santa Ana, California
Heavenly Body	44-30748	N8195H	B-25J	Erickson Aircraft Collection	Madras, Oregon
Hot Gen	45-8883	C-GCWM	B-25J	Canadian Warplane Heritage Museum	Hamilton, Ontario
How Boot That?	44-28925	N7687C	B-25J	Cavanaugh Flight Museum	Addison, Texas
Huaira Bajo (not yet airworthy)	44-31173	LV-GXH	B-25J	Proyecto B-25 Mitchell (Huaira Bajo)	General Rodriguez Airport, General Rodriguez, Argentina
In the Mood	44-29199	N9117Z	B-25J	National Museum of WWII Aviation	Colorado Springs, Colorado
Killer B	44-86697	N62163	B-25J	Tom Reilly Vintage Aircraft Inc.	Wilmington, Delaware
Lady Luck	45-8884	N5833B	B-25J	Pat Harker, Lady Luck LLC.	Blaine, Minnesota
Maid in the Shade	43-35972	N125AZ	B-25J	Comm. Air Force, Arizona Wing	Mesa, Arizona
Miss Hap	40-2168	N2825B	B-25	American Airpower Museum	Farmingdale, New York
Miss Mitchell	44-29869	N27493	B-25J	Comm. Air Force, Minnesota Wing	South St. Paul, Minnesota
Old Glory	44-28938	N7946C	B-25J	John Ward Aviation Co. Inc.	Coulterville, California
Pacific Princess	43-28204	N9856C	B-25J	B-25 Mitchell LLC	Missoula, Montana
Panchito	44-30734	N9079Z	B-25J	Larry Kelley	Georgetown, Delaware
Paper Doll	44-86698	N325N	B-25J	Fagen Fighters WWII Museum	Granite Falls, Minnesota
Photo Fanny	44-30423	N3675G	B-25J	Planes of Fame Air Museum	Chino, California
Pulk (not yet airworthy)	44-31508	N6578D	B-25J	Reevers Warbirds	Adelaide, Australia
Red Bull	44-86893	N6123C	B-25J	The Flying Bulls	Salzburg, Austria
Russian To Get Ya	44-30456	N747AF	B-25J	Lewis Air Legends	San Antonio, Texas
Sarinah	44-29507	PH-XXV	B-25J	Royal Netherlands Air Force Historical Flight	Gilze-Rijen, Netherlands
Semper Fi	44-30988	N5865V	PBJ-1	Comm. Air Force, Southern California Wing	Camarillo, California
Show Me	44-31385	N345TH	B-25J	Comm. Air Force, Missouri Wing	St. Louis, Missouri
Super Rabbit	44-86725	N25NA	B-25J	The Oklahoma Museum of Flying	Bethany, Oklahoma
Take-Off Time	44-30832	N3155G	B-25J	Claire Aviation Inc.	Philadelphia, Pennsylvania
Tondelayo	44-28932	N3476G	B-25J	Collings Foundation	Stow, Massachusetts
Tootsie	44-30606	N201L	B-25J	TSM Enterprises	Carson City, Nevada
Yankee Warrior / Rosie's Reply	43-3634	N3774	B-25D	Yankee Air Museum	Ypsilanti, Michigan
Yellow Rose	43-27868	N25YR	B-25J	Comm. Air Force, Central Texas Wing	San Marcos, Texas

INDEX

BIBLIOGRAPHY

Mitchell Masterpieces Vol. 2
2019, ISBN 978-90-8616-237-6
Wim Nijenhuisn, Lanasta, Emmen

Mitchell Masterpieces Vol. 1
2017, ISBN 978-90-8616-236-9
Wim Nijenhuisn Lanasta, Emmen

Warbird Factory
2015, ISBN 978-0-7603-4816-1
John Fredrickson, Quarto Publishing
Group U.S.A. Inc.

Kansas City B-25 Factory
2014, ISBN 978-1-4671-1197-3
John Fredrickson & John Roper, Arcadia
Publishing

B-25 Factory Times
2013, ISBN 978-90-8616-304-5
Wim Nijenhuism Media Primair Model-
bouw B.V.

B-25 Mitchell Camera Ships
2011, A non-commercial study
Wim Nijenhuis

B-25 Mitchell Catch-22
2011, A non-commercial study
Wim Nijenhuis

B-25 Mitchell Hanover Street
2011, A non-commercial study
Wim Nijenhuis

B-25 Mitchell The Ultimate Look
2008, ISBN 978-0-7643-2930-2
William Wolf, Schiffer Publishing Ltd.

B-25 Mitchell in civil service
1997, ISBN 0-9637543-5-1
Scott A. Thompson, Aero Vintage Books

Fairfax Ghosts
1995, ISBN 0-9647176-0-3
George R. Bauer, National B-25 Preservation
Group

Bombing Twins, Allied Medium Bombers
1991, ISBN 1-855323125
Michael O'Leary, Osprey Aerospace

B-25 Mitchell, The Magnificent Medium
1992, ISBN 0-9625860-5-6
Norman L. Avery, Phalanx Publishing Co.,
Ltd.

Confederate Air Force
1991, ISBN 1-85532-172-6
Peter March, Osprey Aerospace

Bombing Iron
1987, ISBN 0-85045-765-3
Michael O'Leary, Osprey Publishing Limited